TD
892

INDUSTRIAL NOISE CONTROL HANDBOOK

PAUL N. CHEREMISINOFF
and
PETER P. CHEREMISINOFF

CONTRIBUTORS

Ernest E. Allen
E. J. Bonano
A. C. Casciato
N. P. Cheremisinoff
David Marsh
Pandit G. Patil
Anthony J. Schneider
Charles E. Wilson

ANN ARBOR SCIENCE
PUBLISHERS INC
P.O. BOX 1425 • ANN ARBOR, MICH. 48106

Copyright © 1977 by Ann Arbor Science Publishers, Inc.
230 Collingwood, P. O. Box 1425, Ann Arbor, Michigan 48106

Library of Congress Catalog Card No. 76-46023
ISBN   0-250-40144-4

Manufactured in the United States of America
All Rights Reserved

FOREWORD

Controlling noise is an increasing challenge to the industrial world. Reasons for business to be concerned about industrial noise are many, and include the Occupational Safety and Health Act (OSHA 1970). Studies have shown that workers exposed to high noise levels have low working efficiency. Noise can furthermore be responsible for a wide range of psychological and physiological ills such as nervous tension, heart disorders and circulatory problems.

This volume was designed for use by engineers faced with industrial noise problems. It should also be of use to consultants and planners, as well as the student. It is written in general technical language aimed to facilitate its use and stresses the practical rather than the theoretical. The Occupational Safety and Health Act requires that employers provide a noise control program whenever employees work in an environment exposing them to such hazards. Hopefully, this book will provide the necessary information in meeting OSHA requirements, thus making possible a safer and more productive work environment.

The authors are indebted to the many organizations that provided information, photos and data contained herein and have acknowledged them wherever possible in the text. Finally, gratitude is also expressed to the contributing experts who gave of their valuable time, experience and knowledge in the accompanying chapters.

Paul N. Cheremisinoff
Peter P. Cheremisinoff

CONTENTS

1	Introduction	1
2	Noise and Effects on Man	5
3	Noise Legislation	17
4	Acoustics and the Sound Field	41
5	Engineering Controls and Systems Design	55
6	Personal Safety Devices	69
7	Enclosures, Shields and Barriers—Designing With Lead	83
8	Noise Reduction With Glass	143
9	Additional Sound Control Materials	175
10	Silencers and Suppressor Systems	203
11	Fundamentals of Vibration	219
12	Vibration Control Applications	241
13	Abatement and Measurement of Control Valve Noise	273
14	Hydrodynamic Control of Valve Noise	283
15	Ventilating System Noise Control	293
16	Instrumentation for Noise Analysis	305
17	Audiometric Testing and Dosimeters	325
18	Noise Level Interpolation and Mapping	337
19	Glossary	351
	Index	359

CHAPTER 1

INTRODUCTION

Every industry, trade, occupation and process using equipment, methods, apparatus or materials that generate noise above certain levels involves elements of danger to the health and safety of persons employed therein or of persons frequenting such areas. Medical studies and other sources of evidence indicate that noise having certain characteristics affects the auditory system of the human body, causing loss of hearing.

Loss of hearing occurs due to the cumulative effect of exposure to sound above a maximum intensity and over a maximum duration in a given period of time. The basic permissible intensity, as established by the Occupational Safety and Health Administration, is 90 dBA for a duration of 8 hours out of a day.

When employees are subjected to sound exceeding prescribed levels, the employer must institute feasible engineering or administrative controls designed to decrease noise levels in working areas. If the above-mentioned controls fail to reduce noise levels below the prescribed level, personal protective equipment must be provided, and their proper use by the exposed personnel must be enforced by management. To monitor the effectiveness of the engineering and administrative controls, and the proper use and effectiveness of personal protective equipment, a continuing effective hearing conservation program should be administered.

The Department of Labor's Occupational Noise Standards state: "In all cases where the sound levels exceed the values shown herein, a continuing, effective hearing conservation program shall be administered" (Bulletin #334, page 11). The sound level value referred to is 90 dBA for 8 hours per day exposure. Where the sound level above this magnitude cannot be reduced by engineering means, and dependence must be placed on administrative controls or on ear protective equipment, a hearing conservation program is required by law. Personal protective equipment is, however, only to be considered a temporary expedient at best.

The Department of Labor's Occupational Noise Standards (Bulletin #334, revised 1971) specifies three noise control measures and defines them as follows:

Engineering Controls are those which reduce the sound intensity either at the source of the noise or in the hearing zone of the worker.

Administrative Controls are those which limit duration of workers' exposure to noise levels above 90 dBA to the table of Permissible Noise Exposures.

Personal Protective Equipment (ear muffs, plugs, etc.) shall be provided and used to reduce sound levels within the permissible levels. It is important to note that the use of personal protective equipment is considered by the Department to be an interim measure and is not acceptable as a permanent solution to noise problems.

Surveys indicate that more than half of industrial machines generate noise levels of 90 to 100 decibels. This presents a definite health hazard to workers. Noise-induced hearing loss is dangerous because, once incurred, normal hearing can almost never be fully restored.

Four options are open in arresting this problem. The best is to reduce the noise level at its source. However, this is not always feasible. Reduction of noise generated by the machine may be accomplished by:

- proper acoustical design in machinery,
- modifying existing machine design,
- muffling, or
- changing the process entirely.

A second approach is to reduce the amount of sound transmitted through the plant. This can be accomplished by:

- increasing the distance between the work vicinity and source,
- installing acoustical barriers between the work area and sound source, or
- mounting equipment on vibration stands to reduce noise transmission through the building.

This second list may prove to be costly and ineffective on the noise levels in the immediate area.

The third choice is to revise operational procedures. This may involve:

- changing job schedules,
- rotating personnel, or
- providing longer or more frequent work breaks.

This approach cuts down on continuous noise levels; however, it may be uneconomical.

A final approach is to provide the individual with personal ear protection. This seems to be the simplest solution, both economically and to the machine operator; however, there are certain drawbacks. The majority of individuals find personal ear protection devices uncomfortable, irritating and, in many cases, cumbersome.

CHAPTER 2

NOISE AND EFFECTS ON MAN

One of the major problems facing industry today is noise pollution. Data show that over 10% of the U.S. working force suffers some hearing impairment due to exposure to high noise levels. We will examine here the overall effects of noise on an industrial worker, not only in terms of hearing loss, but also work quality.

THE HUMAN EAR

The human ear diagrammed in Figure 2.1 can be broken down into three basic parts. The "outer ear" contains the external auditory canal which functionally carries sound waves to the tympanic membrane, or

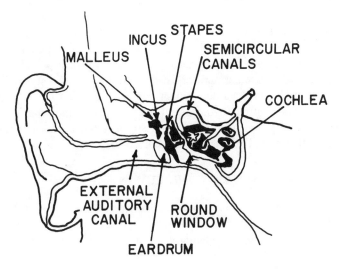

Figure 2.1 Diagram of the human ear.

eardrum. The "middle ear" begins with the eardrum which is a strong flexible tissue with a cone-type construction. When sound waves travel through the external auditory canal and hit the eardrum, they will cause it to vibrate. Also, part of the middle ear is the malleus or hammer, which is a bone that is connected to the eardrum. The malleus is linked to two other bones which move within a very small space. This assemblage of bones, the smallest in the human body, are known as the ossicles and transmit sound waves to the "inner ear" from the eardrum. They also modify sound by either amplifying or diminishing it to protect the inner ear. At the other end of these bones is the foot of the stapes which directly transmits sound to the "inner ear" and is approximately one-thirtieth the area of the eardrum.

We can see that the middle ear regulates the level of sound that enters the external auditory canal, and also protects the inner ear. It is approximately 2 cm^3 in volume, filled with air which dampens the low-frequency rocking of the ossicles. In addition to this arrangement, there are two muscles attached to the stapes and the malleus which are known as the tensor tympani and the stapedius muscles. The function of these muscles is to tighten the eardrum and the motion of the ossicles, thereby lessening the efficiency of sound transmission. This function is known as the acoustic reflex and is carried out by command of the brain just after a very loud sound reaches the eardrum.

The inner ear is a very complex system of boney, fluid-filled crevices lying deep inside the temple. The components which make up the inner ear have nothing to do with hearing but are responsible for our senses. One component is the utricle, which gives us our sense of acceleration and gravity. Another is an arrangement of three semicircular canals which gives us our sense of orientation space and balance. The part of the ear with which we actually hear is called the cochlea. When sound waves travel through the external auditory canal, the foot of the stapes bone knocks against the oval window which is a wide opening in the cochlea, and sound is transmitted to the liquid inside. The round window that lies just below the oval window is an elastic membrane which is the final component that sound reaches in the human ear.

THE EFFECTS OF NOISE ON THE HUMAN EAR

The ear has its own defense mechanism against noise—the acoustic reflex. However, this reflex has vital weak points in its defenses. One is that the muscles within the middle ear can become fatigued and slow if overused. In persons who work in an environment with high noise levels these muscles will gradually lose their strength and thus more noise will reach the inner ear. Secondly, these muscles can be affected by chemicals within the working environment. Finally, the acoustic reflex is an ear-to-brain-to-ear circuit which takes at least nine-thousandths of a second to perform.

Persons with poor acoustic reflex usually are subjected to temporary hearing loss when they come in contact with a loud noise. Much of the temporary hearing loss caused by noise occurs during the first hour of exposure. Recovery of hearing can be complete several hours after the noise stops. In short, the ear will recover to its full hearing potential after its muscles have had time to rest. However, the period of recovery is dependent upon individual variation and the level of noise which caused the deafness.

TOLERANCES

Persons in different age groups have dissimilar tolerances to various noise levels. There have been many investigations made of the threshold of hearing. Figure 2.2 illustrates data indicating the thresholds of hearing and tolerance. The data were obtained by Robinson and Dadson in 1956 upon testing a group of 51 people whose average age was 20. Additional data were obtained by Muson on eight men and two women with the average age of 24.[1]

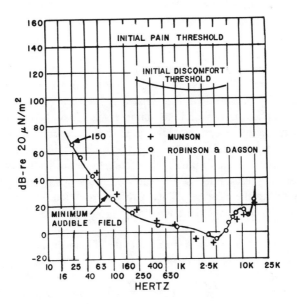

Figure 2.2 Thresholds of hearing and tolerance.[1]

We can see from this data that when young people who have good hearing ability are tested, a characteristic known as the minimum audible field (MAF) is obtained. This characteristic indicates the level of a tone that can just be detected in very quiet surroundings under free field conditions as a function of frequency of the tone. Under free field conditions the sound source is equidistant from any objects including the ground. Therefore, the sound pressure is distributed uniformly in all directions, and doubling the distance from the source will cut the sound pressure in half. The data points shown in Figure 2.2 are accurate to within 5 dB, and this threshold curve indicates that at low frequencies the sound pressure level must be high before a tone can be detected. The graph also shows that at a sound level of about 120 dB a listener would be very uncomfortable. At 140 dB the listener would experience pain.

THE DANGEROUS PROPERTIES OF NOISE

Noises that pose the greatest threat to the human body are those which are the highest pitched, loudest, poorest in tone, and longest lasting. Another dangerous type of sound—capable of rupturing the eardrum—is the sound of an explosion. However, when only the eardrum is ruptured the inner and middle ear are protected. In time, the eardrum heals and full hearing is usually restored.

Deafness due to noise usually occurs in conjunction with a fairly common hearing disorder known as recruitment of loudness. In a person who has this disorder, the zone between what can just be heard and what is too loud is much narrower than normal. Persons with this affliction would have much difficulty in detecting not only weak sounds but sounds which are fully audible to the normal ear. However, the recruitment ear will retain its sensitivity for loud sound levels. Another problem that a person with recruited ears would face is the discomfort of hearing aids. The hearing aid is a microphone that transmits sounds from the surrounding environment to an amplifier connected to a small loudspeaker built into an earplug and aimed at the eardrum. The major problem is that the sounds entering the hearing aid have to be amplified enough to be heard loudly, and at that level the sound may produce discomfort (see Table 2.1).

Researchers have analyzed noise and its effects on the human ear and have come up with several properties of noise which contribute to the loss of hearing. One of these properties is the "overall sound level of the noise spectrum." Here it has been noted that noise whose overall sound level is below 80 dB is reasonably safe. However, many industrial

Table 2.1 Percentage of People with Impaired Hearing in a Noise-Exposed Group

Equivalent Continuous Sound Level (dBA)	Years of Exposure								
	0	5	10	15	20	25	30	35	40	45
80 Risk, %	0	0	0	0	0	0	0	0	0	0
% with impaired hearing	1	2	3	5	7	10	14	21	33	50
85 Risk, %	0	1	3	5	6	7	8	9	10	7
% with impaired hearing	1	3	6	9	13	17	22	30	43	57
90 Risk, %	0	4	10	14	16	16	18	20	21	15
% with impaired hearing	1	6	13	18	22	26	32	41	54	65
95 Risk, %	0	7	17	24	28	29	31	32	29	23
% with impaired hearing	1	9	20	28	34	39	45	53	62	73
100 Risk, %	0	12	29	37	42	43	44	44	41	33
% with impaired hearing	1	14	32	42	48	53	58	65	74	83
105 Risk, %	0	18	42	53	58	60	62	61	54	41
% with impaired hearing	1	20	45	57	64	70	76	82	87	91
110 Risk, %	0	26	55	71	78	78	77	72	62	45
% with impaired hearing	1	28	58	75	84	88	91	93	95	95
115 Risk, %	0	36	71	83	87	84	81	75	64	47
% with impaired hearing	1	38	74	87	93	94	95	96	97	97

Note: Percentage of people with impaired hearing in a nonnoise-exposed group is equal to percentage in a group exposed to continuous sound levels below 80 dBA. Age = 18 years + years of exposure.

noise levels are above 80 dB. To say how these levels would affect the personnel, more specific information on the types of noise would be required.

Another dangerous property of industrial noise is "the shape of the noise spectrum." Various studies have revealed that temporary hearing loss, also known as temporary threshold shift (TTS), is a function of the spectrum of the noise. The ear is specifically sensitive to frequencies above 1 kHz, and in fact most cases of hearing loss occur at these frequencies. Noise containing concentrated energy within the octave bands 600-1200 Hz and higher is much more dangerous to the ear than noise below 600 Hz. Also, a pure tone at a specific level is more dangerous than a band of noise at the same level, both being at the same frequency.

The next dangerous property of noise is "total exposure duration." When exposure to noise permanently lessens a person's sensitivity for hearing, a permanent threshold shift (PTS) has taken place. Through studies of PTS and TTS, it has been determined that the longer a person is exposed to high noise levels, the more his hearing ability will decrease.

However, the hearing loss due to the noise may not continue until the person is totally deaf. Glorig in 1961 determined that permanent deafness at a frequency of 4 kHz from daily exposures of 5 to 8 hours reached a maximum at approximately 12 years of exposure.[1]

A final characteristic of noise that should be mentioned is the temporal distribution of noise. However, energy in noise is distributed across time and its final effect on the threshold shift is a function of total energy. It has been determined that partial noise exposures are related closely to the continuous A-weighted noise level by equal energy amounts. The relation between energy and the amount of exposure can be stated as: twice the energy (3 dBA increase) is acceptable for every halving of exposure time, without any increase in danger. Many studies have revealed that significantly high noise levels can be tolerated if exposure time is decreased sufficiently. Table 2.2 indicates permissible noise exposures according to Department of Labor regulations.

Table 2.2 Permissible Noise Exposures

Duration per Day (hours)	Sound Level dBA Slow Response
8	90
6	92
4	95
3	97
2	100
1½	102
1	105
½	110
¼ or less	115

HOW NOISE CAN AFFECT A WORKER'S MIND AND OUTPUT[1]

Noise affects the mind and changes emotions and behavior in many ways. Most of the time we are unaware that noise is directly affecting our minds. It interferes with our communication, disturbs our sleep and arouses our sense of fear. Psychologically, noise stimulates us to a nervous peak. It is overly-arousing and presents too high a level of stimulation. Too much arousal causes a person to be too ready and, as a result, he will tend to make more mistakes. Experiments in laboratories have revealed that the presence of continuous, loud noises affects the working efficiency in laboratory tasks which require long and concentrated attention. The effects of noise increase the frequency of momentary lapses in efficiency.

Noise has its effects on manual workers as well. These effects can be seen in one case history where a Dr. Jansen of West Germany conducted a study of the psychological effects of noise on the workers of various German steel factories. Over 1000 workers were interviewed and tested. These studies compared two-thirds of the employees who worked in surroundings that had noise levels above 90 dBA to one-third who worked in surroundings less than 90 dBA. The average worker was 41 years old and had been on his job for 11 years. Workers of both groups were matched as closely as possible according to their economic, social and ethnic backgrounds. The results of the comparison revealed that the employees who worked in the quieter surroundings were easier to interview than the employees who worked in the noisier surroundings. Workers exposed to more noise were found to be more distrustful, aggressive and, in some cases, paranoic. It was claimed that the behavior of the latter groups was due in part to their noise-induced loss of hearing. Therefore, noise affects a worker's behavior not only on the job, but also at home. This study also revealed that the workers exposed to higher noise levels had more than twice as many family problems.

Since noise affects a worker's attitude and personality, it also affects his output. It can interefere with his communication greatly; as the noise gets louder, low speech frequencies become more important. To have 90% speech intelligibility between two workers standing about a meter apart, the background noise of mixed frequencies cannot exceed 95 dB, and low-frequency background noise cannot exceed 105 dB (see Figure 2.3).

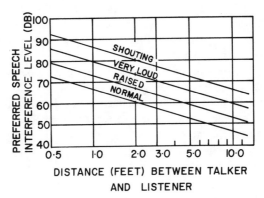

Figure 2.3 Characteristics of speech sound levels in relation to distance.

Noise can cause a decrease in the quality of work output when the background noise exceeds 90 dB. There have been studies, however, which show that the effects of noise on work output are largely dependent on the type of work. For instance, if a person were put on a time-consuming job that requires constant vigilance, noise would be a degenerative factor in his output. High noise levels would tend to cause a higher rate of mistakes and accidents rather than a direct slowdown of production. Results show that a worker's attention to the job at hand will tend to drift as noise levels increase.

EFFECTS OF NOISE ON THE HUMAN BODY

The first studies on the effects of noise on the human body were conducted by Dr. G. Lehmann, Director of the Max Planck Institute, in the mid-1950s. His team of researchers had determined that noise has an explicit effect on the blood vessels, and especially the smaller ones known as precapillaries. Overall, noise makes these blood vessels narrower. It was also found that noise causes significant reductions in the blood supply to various parts of the body. This fact was discovered through a test which employed transillumination, a technique of shining a light inside the mouth of a patient and observing the redness of the cheeks on the outside. If the cheeks were red, sufficient blood was being supplied; if they were pale, there was a deficient amount of blood. When noise was administered around the patient, his cheeks became more pale, showing that an insufficient amount of blood was being supplied.

Tests were also conducted employing a ballistocardiogram, which is used to measure the record of the heart with each beat. When the test was conducted on a patient in noisy surroundings, it was found that there was a decrease in the stroke volume of the heart. The **results** of these findings lead to one conclusion—that noise at all levels causes the peripheral blood vessels in the toes, fingers, skin and abdominal organs to constrict, thereby decreasing the amount of blood normally supplied to these areas. This shrinking of the small blood vessel is known as vasoconstriction and is a reflex action generated by the nervous system. It is triggered by various body chemicals, namely adrenaline, which is produced when the body is under stress. Later experiments involving finger pulse amplitude produced similar results. These experiments revealed that 3 seconds at 87-dB noise constricted arterioles in the fingers and cut down the volume of blood by one-half. After the noise stopped, it took approximately 5 minutes for the arterioles to fully recover.

Earlier investigations revealed the effects of noise on the blood vessels which feed the brain. These experiments were carried out employing a reograph, which measures the flow of blood in the vessels leading to the brain, and concluded that these vessels dilate in the presence of noise. This is the reason why headaches result from listening to persistent noise. Other studies have shown that noise can induce heart attacks. When the small blood vessels are subjected to noise there is an aggregation of red blood cells within them, and the vessels contract in spasm. Thus, noise will actually cause the blood to thicken and clot and could very well lead to a heart attack.

One final part of the body affected by noise is the nervous system. Noise wears down the nervous system, breaks down our natural resistance to disease and our natural recovery, thus lowering the quality of general health.

AUDIOGRAMS

An important step toward hearing conservation in industry is the monitoring of the hearing ability of employees exposed to noise conditions. The monitoring of hearing is known as audiometry. Through audiometry it is possible to detect the least-intense sound that can be heard (absolute threshold), and the minimal noticeable difference between two sounds (differential threshold), by an employee.

An audiogram is the record of an employee's hearing sensitivity for absolute and differential thresholds as a number of pure-tone frequencies. Hearing sensitivity is measured in terms of deviation in dB found in normal hearing. Normal hearing is defined by the International Standard Organization and the American National Standards Institute, and is illustrated by the absolute threshold levels shown in Table 2.3.

Table 2.3 Pure-Tone Reference Threshold Levels[a]

Frequency (Hz)	dB (re $20\ \mu N/m^2$)
125	45
250	25.5
500	11.5
1000	7
1500	6.5
2000	9
3000	10
4000	9.5
6000	15.5
8000	13

[a] 1964 ISO/1969 ANSI, based on measurements made on National Bureau of Standards 9-A Coupler and Telephonics TDH-39 Earphone fitted with MX-41/AR Cushion. Similar values have been developed for other sources.

14 INDUSTRIAL NOISE CONTROL HANDBOOK

An audiogram is obtained through an audiometer. This device allows pure tones at specific frequencies and intensities to be tested on the employee. There are many types of audiometers used by industry. They can provide frequencies of 500, 1000, 2000, 3000, 4000 and 6000 Hz, with output levels more than 70 dB above the standard threshold level.

THE EFFECTS OF VIBRATION ON MAN

Vibration, like noise, also has harmful effects on the human body. We experience the discomfort of vibration while traveling in boats, cars and planes. Many types of tools and machinery with which we have physical contact vibrate and affect our efficiency in using them. Several physical problems directly related to vibration are listed in Table 2.4.

Table 2.4 Vibration-Related Problems

Effect on Man	Other
Injury	Excessive wear
Fatigue	Excessive noise
Annoyance	Inadequate performance
Interference with performance	Failure to satisfy vibration specifications
Mechanical Failure	
Excessive stress	
Fatigue	
Destructive impacts	

Studies on the effects of vibration have revealed that sensations due to it are not centralized, since it can be felt throughout the entire body. Figure 2.4 illustrates the subjective response of the human body to vibration as a function of frequency. There is a great sensitivity to vibration which occurs in the frequency range of 4 to 10 Hz. Within this range the vibration causes resonance effects of the internal organs and their supports, the upper torso and the shoulder-girdle structures. Many other resonance effects in the body occur also, but these depend on the mode of excitation and the area to which the vibration is applied. Industrial workers who use certain types of hand-held power tools for extended periods may exhibit Raynaud's phenomenon, whereby the fingers become white and numb as when a person is chilled.

NOISE AND EFFECTS ON MAN 15

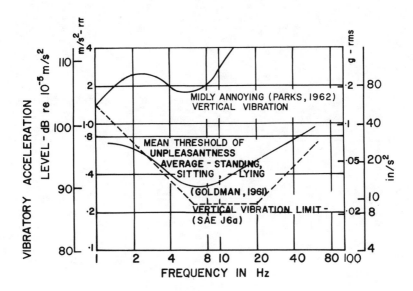

Figure 2.4 Response of the human body to vibratory motion as a function of frequency.[1]

REFERENCE

1. Gross, E. E., Jr. *Handbook of Noise Measurement* (Concord, Massachusetts: General Radio Company, 1974).

CHAPTER 3

NOISE LEGISLATION

OCCUPATIONAL SAFETY AND HEALTH ACT (OSHA)

The Occupational Safety and Health Act was signed on December 29, 1970 and went into effect April 28, 1971. The purpose of this Act is "to assure so far as possible every working man and woman in the nation safe and healthful working conditions and to preserve our human resources." This Act applies to every employer in the country associated with a business that affects commerce. OSHA does not apply to such working conditions that are protected by other federal occupational safety and health laws such as the Federal Coal Mine Health and Safety Act, The Atomic Energy Act, The Metal and Nonmetallic Mines Safety and Health Standards, and the Open Pit and Quarries Safety and Health Standards. This Act puts all state and federal occupational safety and health enforcement programs under federal control with the goal of establishing more uniform standards, regulations and codes with stricter enforcement.

Several of the major aspects of the Act will maintain federal supervision of state programs to obtain more uniform state inspection under federal standards. OSHA will also make it mandatory for employers to keep accurate records of employee exposures to harmful agents which are required by safety and health standards to be monitored and measured. The law provides procedures in investigating violations by delivering citations and monetary penalties upon the request of an employee. OSHA establishes a National Institute of Occupational Safety and Health (NIOSH) whose members have the same powers of inspection as members of OSHA. The Act also delegates to the Secretary of Labor the power to issue safety and health regulations and standards enforceable by law. This last provision is carried out through the Occupational Safety and Health Administration.

OSHA has made many major industries coordinate and revise industrial hygiene, medical and safety activities with all management functions. With this law in effect, companies have to fully understand their responsibilities in maintaining safe working conditions and the rights of employees. Management decisions on process design, equipment selection, job description and material use must be made in consideration of restrictive standards. In the field of employer-employee relations, employers receive demands such as:

- Hazard pay for hazardous jobs and for working time lost due to hazards.
- A union examiner to make tests and evaluate the conditions of the working environment.
- Complete current data on exposure to noise levels, air contaminants and radiation. Also, industrial hygiene test data on the various levels of exposure.
- More thorough decisions on the size of crews and the length of break periods.
- Complaints of hazardous jobs based primarily on visual observations and sensory response rather than actual data taken from a test situation.
- Physical examinations of employees by a physician of their own choice.

However, all the above demands and complaints must be supported by valid test data obtained from periodic monitoring of the work areas by authorized personnel such as industrial hygienists from headquarters environmental health or plant personnel directed by corporate technical officials.

Responsibilities of Employers Under OSHA

OSHA enforces two basic duties which must be carried out by employers: first, to provide each employee with a working environment free of recognized hazards that cause or have the potential to cause physical harm or death; second, to fully comply with the Occupational Safety and Health Standards under the Act.

To carry out the first duty, employers must have proper instrumentation for the evaluation of test data by an expert in the area of industrial hygiene. This instrumentation must be obtained because the presence of health hazards cannot be evaluated by visual inspection. This duty can be used by the employees to allege a hazardous working situation without any requirement of expert judgment. Evaluation of case histories has revealed that the majority of workers' complaints to OSHA relate to health matters. This duty also provides the employer with substantial evidence to disprove invalid complaints.

Responsibilities of Employees Under OSHA

It states within the Act that "each employee shall comply with Occupational Safety and Health Standards and all Rules, Regulations and Orders issued pursuant to this Act which is applicable to his own actions and conduct." This law gives employers the right to take full disciplinary action against those employees who violate safe practices in working methods. An employer can also be cited, however, if a compliance officer observes violation of the Act by an employee instituting unsafe practices.

Administration

The Secretary of Labor controls the full administration and enforcement of OSHA, and has set up a separate agency called the Occupational Safety and Health Administration (OSHA) and the Occupational Safety and Health Review Commission (OSHRC). The Act also sets up a National Institute of Occupational Safety and Health (NIOSH). NIOSH is an agency which undertakes research programs to obtain data by which standards are revised. OSHA sets the standards and acts as an enforcement agency, and OSHRC is an appeals agency.

NIOSH has the function of undertaking research and some educational functions assigned to the Secretary of Health, Education and Welfare. NIOSH also develops recommended occupational safety and health standards. The organization of NIOSH is illustrated in Figure 3.1.

The Occupational Safety and Health Administration (OSHA) has full responsibility for administering the Occupational Safety and Health Act. This administration has the full authority to enter and inspect the plant of any employer, to evaluate all working conditions and to issue any citations or penalties for violations of the standards. An Assistant Secretary of Labor is the administrative officer. Figure 3.2 illustrates the organizational structure of OSHA with all of its offices.

OSHA Standards

There are several different types of standards under this Act, one of which is known as a *Permanent Standard*. The proposed standard is published in the *Federal Register* so that people have an opportunity to review it and send in written comments and data on it. Any person may file objections to standards and may also call for a public hearing. OSHA has the authority to officially publish permanent standards and revoke or revise early standards on its own basis or on the basis of information submitted by: an employer or employee organization; a nationally recognized

20 INDUSTRIAL NOISE CONTROL HANDBOOK

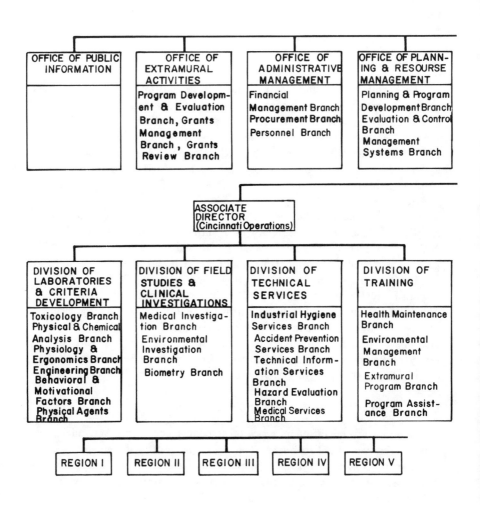

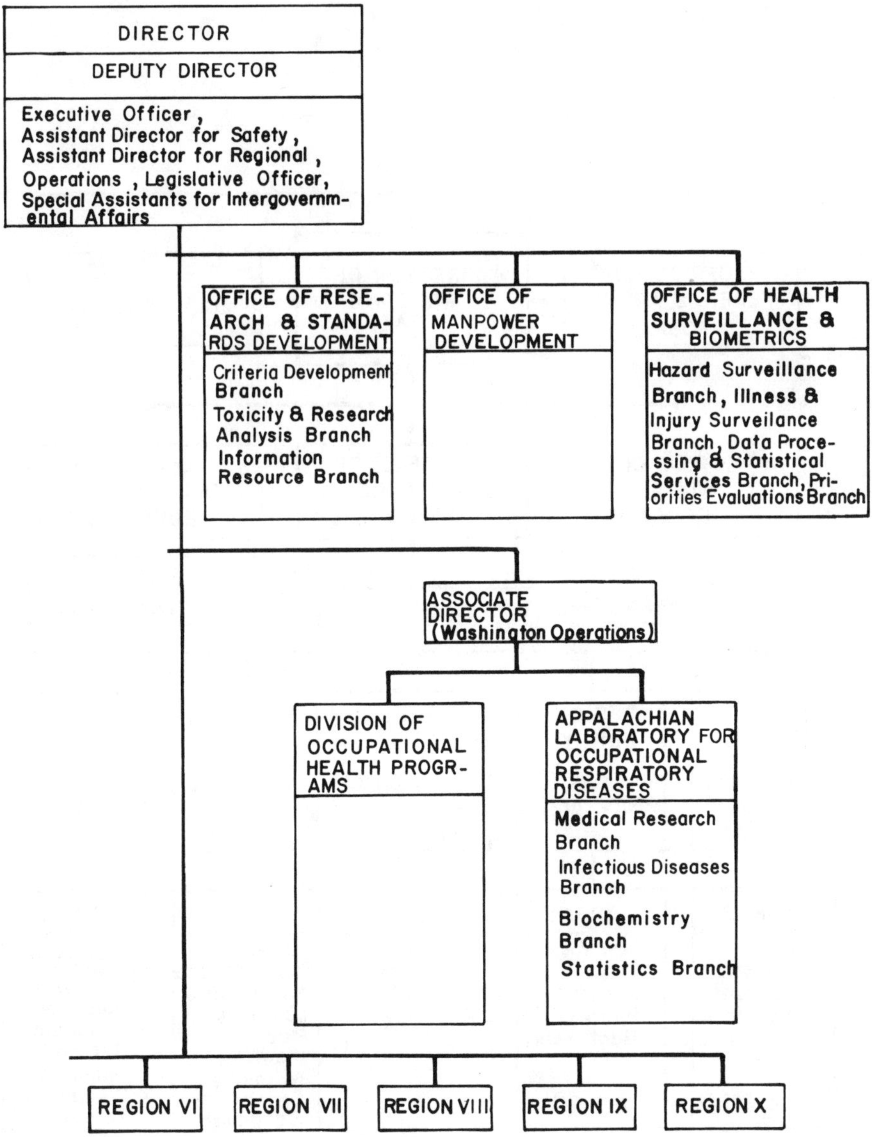

NOISE LEGISLATION 21

```
┌─────────────────────────────┐
│         DIRECTOR            │
├─────────────────────────────┤
│      DEPUTY DIRECTOR        │
├─────────────────────────────┤
│ Executive Officer,          │
│ Assistant Director for Safety, │
│ Assistant Director for Regional, │
│ Operations, Legislative Officer, │
│ Special Assistants for Intergovernm- │
│ ental Affairs               │
└─────────────────────────────┘
```

OFFICE OF RESEARCH & STANDARDS DEVELOPMENT	OFFICE OF MANPOWER DEVELOPMENT	OFFICE OF HEALTH SURVEILLANCE & BIOMETRICS
Criteria Development Branch Toxicity & Research Analysis Branch Information Resource Branch		Hazard Surveillance Branch, Illness & Injury Surveillance Branch, Data Processing & Statistical Services Branch, Priorities Evaluations Branch

ASSOCIATE DIRECTOR (Washington Operations)

DIVISION OF OCCUPATIONAL HEALTH PROGRAMS	APPALACHIAN LABORATORY FOR OCCUPATIONAL RESPIRATORY DISEASES
	Medical Research Branch Infectious Diseases Branch Biochemistry Branch Statistics Branch

| REGION VI | REGION VII | REGION VIII | REGION IX | REGION X |

Figure 3.1 Organization of National Institute for Occupational Safety and Health.

INDUSTRIAL NOISE CONTROL HANDBOOK

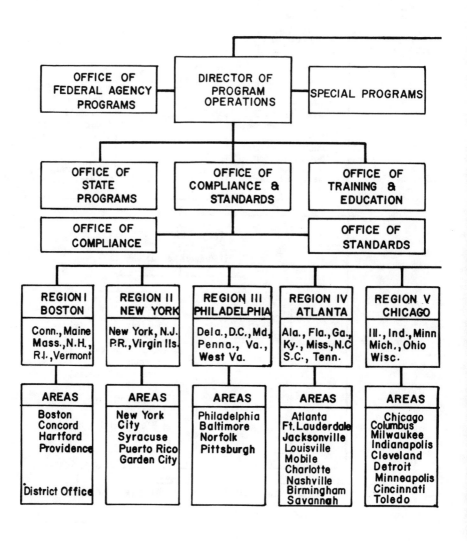

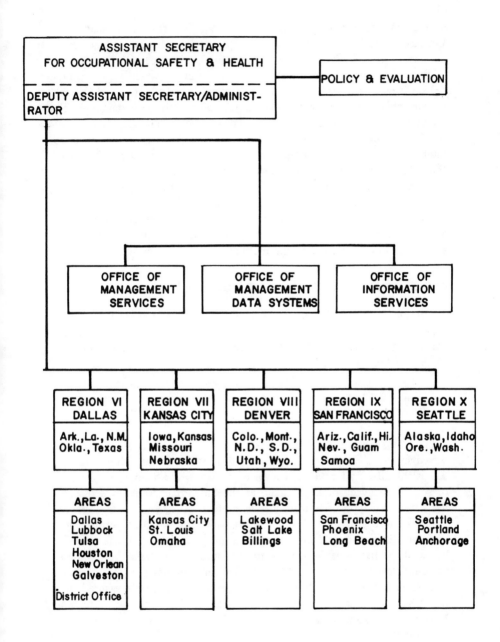

Figure 3.2 Organization of Occupational Safety and Health Administration.

standards-producing organization; NIOSH; the Department of Health, Education and Welfare; and a state or political subdivision. Recommendations from such groups must be made within 270 days.

Basically, there are four distinct categories under which OSHA standards fall. Category number one contains *Vertical Standards*—those that apply to particular industries with specifications that apply to certain operations. Category number two covers *Horizontal Standards*—standards that apply to all work places and relate to very wide areas such as sanitation and illumination regulations. Category number three covers *Design Standards*—those that provide an OSHA inspector with criteria to cite an employer for violations in his production system. The fourth category covers *Performance Standards*—threshold limit values (TLV's) established by the American Conference of Governmental Industrial Hygienists. Essentially, a performance standard states the level to which a plant must raise its safety conditions and leaves the method up to the employer.

OSHA'S REGULATIONS ON NOISE CONTROL

Section 50-204.10 of the Act establishes acceptable noise levels and exposures for safe working conditions, and gives various means of actions which must be taken if these levels are exceeded. Below, paragraph (a) of this section is quoted:

> Protection against the effects of noise exposure shall be provided when the sound levels exceed those shown in Table I of this Section when measured on the A scale of a standard sound level meter at slow response. . . .

Table I of Section 50-204.10 is illustrated in Table 3.1. As can be observed in the table the acceptable sound intensity for an 8-hour working day is 90 dBA. At a 90-dBA level of exposure for 8 hours the amount of sound energy absorbed is taken as the limit of exposure which will not cause any type of hearing loss in more than 20% of those exposed. Table 3.1 also gives higher sound levels with corresponding permissible exposure times that will not induce any more damaging effects than 8 hours of 90-dBA noise. Workers in any industry must not be exposed to sound levels greater than 115 dBA for any amount of time.

Paragraph (a) of this Section says noise levels must be measured on the A scale of a standard sound level meter at slow response. The sound level meter is a measuring device which indicates sound intensity. The A scale on this instrument is one of many, and when operating on this scale the meter functions in a manner very similar to the human ear; it is less responsive to low pitched tones. "Slow response" is a particular setting on the meter, and when the meter is at this setting it will average out high-level noises of short-lived duration.

Table 3.1 Permissible Noise Exposures[a]

Duration per Day (hours)	Sound Level dBA Slow
8	90
6	92
4	95
3	97
2	100
1½	102
1	105
½	110
¼ or less	115

[a] When the daily noise exposure is composed of two or more periods of noise exposure at different levels, their combined effect should be considered, rather than the individual effect of each. If the sum of the following fractions: $C_1/T_1 + C_2/T_2 \ldots C_n/T_n$ exceeds unity, then the mixed exposure should be considered to exceed the limit value. C_n indicates the total time of exposure at a specified noise level, and T_n indicates the total time of exposure permitted at that level.

Impulse or Impact Noise

A quote from paragraph (d) of Section 50-204.10 of OSHA states, "Exposure to impulsive or impact noise should not exceed 140 dB peak sound pressure level." This establishes the maximum sound level to which an employee is to be exposed, regardless of how minute the time span. In Table 3.1 the maximum permissible level for steady noise was 115 dBA. For impact noise a higher level of 140 dB is acceptable because the noise impulse due to impacts (such as explosions) is over before the human ear has time to fully react to it. Levels of impact noise are measured by impact meters or oscilloscopes.

Variable Noise

Regulations of variable noise levels are covered under paragraph (c) of Section 50-204.10 in OSHA. This paragraph states: "If the variations in noise levels in value maxima at intervals of 1 second or less, it is considered to be continuous." Therefore, when the level on the meter goes from a relatively steady reading to a higher reading, at intervals of one second or less, the higher reading is taken as the continuous sound level.

Sounds of short duration occurring at intervals greater than 1 second should be measured in intensity and duration over the total work day. These data should be entered into the equation $C_1/T_1 + C_2/T_2 \ldots Cu/Tr$ to obtain the allowable limit. Such sounds may be analyzed using a sound level meter and should not be treated as impact sounds.

Engineering Control Measures

Control measures are covered in paragraph (b) of Section 50-204.10 which says: "When employees are subjected to sounds exceeding those listed in Table I of this Section, feasible administrative or engineering controls shall be utilized. . . ." Below are several samples of engineering control measures that reduce noise levels at the source or in the hearing area of the employees.

A. **Substitution of Machines**
   1. Larger, slower machines for smaller, faster ones
   2. Step dies for single operation dies
   3. Presses for hammers
   4. Rotating shears for square shears
   5. Hydraulic for mechanical presses
   6. Belt drives for gears

B. **Maintenance of Equipment**
   1. Replacement or adjustment of worn and loose or unbalanced parts of machines
   2. Lubrication of machine parts and use of cutting oils
   3. Properly shaped and sharpened cutting tools

C. **Substitution of Processes and Techniques**
   1. Compression for impact riveting
   2. Welding for riveting
   3. Hot for cold working
   4. Pressing for rolling or forging

D. **Dampening Vibration in Equipment**
   1. Increase mass
   2. Increase stiffness
   3. Use rubber or plastic bumpers or cushions (pads)
   4. Change size to change resonance frequency

E. **Reducing Sound Transmission through Solid Materials**
   1. Flexible mounts
   2. Flexible sections in pipe runs
   3. Flexible shaft couplings
   4. Fabric sections in ducts
   5. Resilient flooring

F. **Include Noise Level Specifications when Ordering New Equipment**

G. **Reducing Sound Produced by Fluid Flow**
   1. Intake and exhaust mufflers
   2. Fan blades designed to reduce turbulence
   3. Large, low-speed fans for smaller, high-speed fans

H. **Isolating Operator**
   1. Provide a relatively sound-proof booth for the operator or attendant of one or more machines.

I. Isolating Noise Sources
   1. Completely enclose individual machines
   2. Use baffles
   3. Confine high-noise machines to insulated rooms

The various controls listed above can be achieved with little expense or effort to plant personnel. The Department of Labor expects employers to investigate the possibilities of controlling noise through engineering methods.

Administrative Controls

Sometimes engineering controls are not enough to maintain permissible noise levels. In such cases, administrative controls should be employed to help combat the problem. Several measures which could be taken by the administration are listed below.

1. See that workers who have reached the maximum limit of duration for a high noise level, in accordance with Table 3.1, spend the remainder of the day working in an environment with a noise level well below 90 dBA.
2. Arrange work schedules so that employees working most of the day at the 90-dBA limit are not exposed to any higher noise levels.
3. Where the man-hours required for a job exceed the permissible time for one man in a day for the prevailing sound level, divide the work among as many men as needed, either successively or together, to keep individual noise exposure within the permissible time limits.
4. Perform necessary high-level noise-producing operations during the night or at other times when there is a minimum number of employees exposed.
5. If less than full-time production of a high-noise-level piece of equipment is needed, arrange to run it at a portion of each day, rather than all day, for only a few days of the week.

WALSH-HEALEY PUBLIC CONTRACTS ACT

The Federal Walsh-Healey Public Contracts Act regulations took effect on May 20, 1969. To comply with Walsh-Healey regulations on industrial noise exposure, an industry must measure the noise level of its working environment. This type of survey will provide valuable data with which an inspector can evaluate working conditions. The data may be obtained by sound survey meters with A-, B- and C-weighted filters. To test for compliance only the A-weighted measurements are needed. The C-weighted filter has an unweighted flat response between 40 and 8000 Hz, and can provide additional measurements for a more complete survey. Therefore, any measurement will weigh all the frequencies equally in that range.

Table 3.1 listed the acceptable set of levels under the Walsh-Healey Act. Also, the band frequencies are the same for OSHA and Walsh-Healey. When industrial levels exceed permissible limits, the Walsh-Healey Act states that a hearing conservation program is mandatory. Steps taken to institute such a program are listed below.

1. Group the processes according to noise output; then group the noisiest processes together. This will minimize the number of people exposed.
2. Be selective when purchasing machinery and specify maximum allowable noise output. If no better method is available, cite the Walsh-Healey limits and require the vendor to meet them.
3. Inform maintenance personnel that noise is a by-product of worn or maladjusted equipment.
4. Inspect the operation for noisy machinery to determine which units can operate as well in a noise-proof enclosure.
5. If noise levels still exceed the regulation, require personnel to use ear protection devices.

The U.S. Department of Labor will cooperate with the employers, whenever and however appropriate, to assure the cooperation of employees.

Walsh-Healey applies to plants with governmental contracts totalling $10,000. In multiplant companies, an inspector may not inspect plants which have no connection with government contracts; but if any one plant is found in violation, the entire company may be penalized. Maximum penalty for a convicted violation is removal from the federal government's approved bidder list for three years. It should be emphasized again that interested parties should obtain interpretation of their own particular case from legal counsel or from the Department of Labor.

PRIMER ON OSHA

A general guide to the Occupational Safety and Health Act of 1970 is herein provided.

Purpose of the Act

All employees of the nation are to be assured healthful and safe working conditions. To accomplish this purpose, mandatory occupational health and safety standards are promulgated. There are also provisions for research, information, education and training in occupational safety and health. The training and education portions to date have consisted of seminars to familiarize industry with OSHA.

Businesses Affected

The law applies to every employer. This law also applies to construction firms unless the firm is working under the Construction Safety Act due to government contract financing. Standards under the Construction Safety Act have been included as part of OSHA standards. An Executive Order by the President has directed each federal agency to draw up safety programs that will in effect be in tune with OSHA. State and municipal employees are excluded from coverage. The Act does not apply to working conditions protected under other federal occupational safety and health laws—such as the Federal Coal Mine Health and Safety Act and Atomic Energy Act of 1954.

Duties of Employers

Each employer shall furnish each employee employment and a place of employment free from recognized hazards causing or likely to cause death or serious physical harm. Other duties are as follows:

1. Initiate such programs as may be necessary to comply with the Act.
2. Permit only those employees qualified by training and experience to operate equipment and machinery.
3. Provide and train employees in the use of appropriate personal protective equipment where necessary.
4. Make all operating departments aware of requirements under the law. Purchasing departments must require that all equipment and materials meet or exceed recognized good safe practice—the minimum of which is that applicable under OSHA standards. Plant engineers, architectural departments, construction and industrial engineers should become familiar with good safe practice, OSHA standards, and any consensus standards such as ANSI, ASME or NFPA standards.
5. Maintain a liaison with the OSHA office in your region and safety officials of your state government.
6. Set specific safety goals for your organization and measure progress toward such goals on a regular basis.

Duties and Rights of Employees

Each employee shall comply with the safety and health standards, and all rules, regulations and orders issued pursuant to the Act which are applicable to his own actions and conduct. Employees have the right to:

1. Request an inspection if they believe an imminent danger exists or a violation of a standard threatens physical harm.
2. Have a representative accompany a compliance officer during the inspection of any work place.
3. Advise a compliance officer of any violation of the Act which is believed to exist in the work place, and to question and be questioned privately by a compliance officer.
4. Have regulations posted to inform them of the protection afforded by the Act.
5. Have locations monitored in order to measure exposure to toxic materials, have access to the records of such monitoring or measuring, and have a record of their own personal exposure.
6. Be advised of any order issued by the Secretary of Labor granting a variance to an employer, be informed when an application for variance is requested, and petition the Secretary for a hearing pertaining to any variance requested.
7. Request the employer to review copies of regulations and standards.
8. Be given medical examinations or other tests to determine whether their health is being affected by an exposure, and have the results of examinations or tests furnished to their physicians.
9. Be informed of citations made to the employer through the requirement that citations be promptly posted, contest the length of time fixed for abatement of the violation on the basis that it is unreasonable, and participate in hearings concerning abatement time.

It is suggested that all levels of management be informed of the duties and rights granted to employees under the Occupational Safety and Health Act.

Occupational Safety and Health Standards

Employers are obligated to familiarize themselves with those standards which apply to them. It should be pointed out that OSHA standards are minimum requirements. The employer must be familiar with or have access to any of the consensus standards that may apply to his operation.

Interim Standards. The National Consensus Standards (ANSI and NFPA) are those on which interested parties have reached substantial agreement after considering diverse views—plus existing federal standards.

Emergency Standards. The Secretary of Labor determines standards necessary to protect employees exposed to toxic substances from physical harm or from new hazards. Once published the Secretary must follow by setting permanent standards.

Permanent Standards. These standards are set by the Secretary of Labor and most adequately assure, to the extent feasible and on the basis of the best available evidence, that no employee will suffer material impairment of health or functional capacity even if regularly exposed for his working life.

Administration

Administration and enforcement is vested in the Secretary of Labor and the Occupational Safety and Health Review Commission. Research is vested in the Secretary of Health, Education and Welfare and will be carried out by the National Institute for Occupational Safety and Health.

Complaints of Violations

Any employee who believes that a violation of a standard or any provision of the Act or any unsafe or unhealthy condition exists may request an inspection by (a) a signed written notice to the Department of Labor, or (b) by an outline of the grounds for notice. A copy of the employee's request must be provided to the employer no later than the time of inspection. The copy need not include the employee's name. If the Secretary of Labor finds no reasonable grounds and no citation is issued, complainants must be notified in writing of final disposition. The Secretary is required to set up **procedures** for informal review where a citation is not issued.

Inspections

Compliance officers may enter upon presenting appropriate credentials to the owner, operator or agent in charge. They are authorized to enter without delay, at any reasonable time, to inspect and investigate during regular working hours. They may question privately an employer, owner, operator, agent or employee. The employer and employee representative may accompany the compliance officer to aid in the inspection. Any person giving advance notice of impending inspection is liable for a $1000 fine, six months imprisonment, or both. It is a federal crime to kill, assault or resist personnel carrying out their duties. Trade secrets are to be treated confidentially.

Upon completion of an inspection, the compliance officer is to confer with the employer to advise of violations found. A citation will not be issued at this time. It will be issued by the area director via registered mail. Reasonable time is fixed for abatement. Each citation must be posted at or near the place of violation. Time for abatement is listed on serious and nonserious violations. No citation can be issued after the expiration of six months following the occurrence of that violation.

Notification of Proposed Penalty

An employer has 15 working days to notify the Department of Labor that he wishes to contest a citation or the penalty, or request a variance. If he fails, the citation and penalty are final. The review commission will, upon review, affirm, modify or vacate the citation or penalty. Review of commission orders can be appealed to the U.S. Court of Appeals. If the penalty is contested, it is not final until reviewed and affirmed by the commission.

Contesting a Citation

Notification must be made to the area director in charge of the local office of the U.S. Department of Labor which initiated the action being contested. The time allowance is 15 working days from receipt of a letter by certified mail containing the Department's proposed penalty. This notification is the "Notice of Contest" and when placed in the mail, it becomes Act I-Scene I of the case which will thereafter go before the Occupational Safety and Health Review Commission. If any of the employees who work at the site of the alleged violation are unionized, a copy of the Notice of Contest must be served upon their union. If any of the employees who work at the site of the alleged violation are not represented by a union, copies of the Notice of Contest must either be posted in a place where employees will see it (the Labor Department's regulation for posting of citations) or be personally served upon such employees. There is no particular form prescribed for the Notice of Contest, but it must clearly identify the citation, notification of proposed penalty, or notification of failure to correct violation which forms the basis for its filing. A Notice of Contest must also contain a listing of the names and addresses of those parties to the case who are caused to be personally served with the same, plus the address of the posting when one is required.

Time for Abatement of Hazards

A citation prescribes a reasonable time for compliance. The time limit may also be contested within 15 working days. The time limit does not begin until there is a final order of the review commission.

Employees also may object within 15 working days to the time fixed in the citation on the grounds of alleged unreasonableness, but only if the employer does not contest. The same review procedures apply as before.

Variances

Any employer can ask the Secretary of Labor for a variance order if he cannot comply with a standard prescribed by the law. There are three types of variance:

1. *Temporary*—when an employer needs more personnel or larger facilities in order to comply with the standard.
2. *Equally Safe Working Conditions*—when an employer can provide just as safe and healthful working conditions as if he complied with the standard.
3. *Safety and Health Experiment*—when an employer takes part in a worker safety or health experiment with approval from the Secretary of Health, Education and Welfare.

Failure to Correct Violations within Allowed Time

If an employer fails, the Secretary will notify him by certified mail of his failure and the assessed penalties. This notification will be final unless contested within 15 days. In good faith—if the employer cannot comply due to factors beyond his control—a hearing will be afforded to affirm or modify the time requirement.

Penalties for Violations

Penalties for violations are as follows: those not corrected in time—up to $1000 per day per violation; for willful or repeated violations—up to $10,000 per violation; for willful violations resulting in death—$10,000 or six months imprisonment (a second conviction doubles this penalty).

Recordkeeping Requirements

OSHA 100—mandatory; OSHA 101—optional, other reports may be used if they contain all information required; OSHA 102—mandatory.

Statistics

The Secretary of Labor and Secretary of Health, Education and Welfare are required to develop and maintain effective programs for collection, compilation and analysis on work injuries and illnesses.

Imminent Dangers

In the event of conditions or practices, so determined, which might cause death or serious physical harm, a court order can be sought to restrain normal operations by shutdown of operations or plant, or by removal of employees from the affected area.

Protection Against Harrassment

No person shall be discharged or discriminated against because he exercises any right under the Act. If so, he may file a complaint within 30 days of the illegal action with the Secretary of Labor. The Secretary of Labor takes the issue to U.S. District Court for appropriate relief.

What an Employer Should be Doing

To assure efficient and orderly organization relating to existing practices and OSHA, a list of priorities to assure compliance with OSHA, while at the same time maintaining an overall program, should be established. Procedures and practices should be reviewed in establishing a plan of action and priorities.

1. Formulate a policy statement outlining duties and responsibilities to be included as a part of the company's program.
2. **Make** certain appropriate personnel in the organization are knowledgeable about OSHA and are kept up-to-date on new developments.
3. Establish an OSHA-required recordkeeping system as outlined in the Department of Labor pamphlet entitled "Record-Keeping Requirements under the Williams-Stieger Occupational Safety and Health Act of 1970."
4. **Analyze** and, where necessary, strengthen safety and health programs.
5. Establish procedures for handling a citation should one be received.
6. Improve liaison with insurance brokers and/or carrier loss prevention personnel. These sources usually can offer expertise in safety and health matters.

7. Establish hazard control priorities through surveys of premises, equipment and materials, based on applicable safety and health standards.

GUIDE FOR IMPLEMENTATION

Policy

The policy of the manufacturing organization should be to provide safe and healthful working conditions in all activities. Consistent with that policy, all managers within an organization are responsible for compliance with OSHA as an integral part of their safety responsibility.

Procedures

Each manager shall assume the responsibility for applying the provisions and standards of OSHA and shall specifically designate two alternates, in a specific order, to serve as the persons responsible in his absence. Under the direction of the manager, there shall be:

1. A specifically established safety program, in writing, which produces results.
2. An active safety committee.
3. A prompt response to recommendations submitted relating to safety or fire prevention.
4. A thorough and effective accident investigation and reporting procedure.
5. A training program for employees and supervisory personnel directly related to possible injury or illness in the location's operations.
6. A specific audit or survey of all premises, equipment and material so that recommendations can be developed to obtain compliance with established standards.

Recommendations may be categorized as follows:

- Immediate attention
- Imminent dangers
- Those items only requiring modest expenditures of time or money
- As schedule permits—those items which are temporarily acceptable but should be corrected
- Preplanning necessary—recommendations requiring new programs, equipment or processes.

It is recognized that some technical violations, which do not constitute hazards, may be attended to during major renovation or new planning. Action on all of the above should be well documented as evidence of good faith. Immediate attention is required to any complaints on the part of employees concerning a possible injury or illness potential. Also necessary is a communication system through which all personnel are kept abreast of new standards or procedures as they are published by the Department of Labor. Specific goals are to be established for the safety program and progress measured toward those goals on a regular basis.

Should there be an inspection by a Department of Labor representative, the inspector should be immediately referred to those management personnel designated as responsible for compliance with the Act, but preferably the location manager. Examine the inspector's credentials to assure that he is a representative of the Department of Labor. Treat the inspector with the respect of his office. It is important to recognize that, by law, Department of Labor inspectors have the right to make an inspection, usually unannounced.

Previous to the inspection, attempt to review with the inspector: company policy concerning the Act; safety program organization and procedures; evidence of safety activities such as minutes of safety meetings, reports submitted after plant inspections, educational activities, noise control, etc.; accident data to emphasize the thoroughness of investigations and the procedures for taking corrective action after an accident occurs. Arrange for the availability of employees who have a right to accompany the inspector during his inspection. Management personnel should make themselves available for discussions with the inspector during and after his plant tour. An exit interview should be insisted upon and a record made of all that the inspector says, particularly if citations are to involve conditions that would either be difficult to correct or require sizeable expenditure. It is important to understand that there is but a 15-day response period to citations after they are received. Upon receipt of the inspector's report, immediately communicate with appropriate headquarters personnel to establish a plan of action in response to recommendations and possible citations.

Recordkeeping Requirements

Each location must be provided a copy of the U.S. Department of Labor publication "Recordkeeping Requirements under the Williams-Stieger Occupational Safety and Health Act of 1970." Forms required are: OSHA 100—Log of Occupational Safety and Health Injuries and Illnesses; OSHA 101—Supplementary Record of Occupational Injuries and Illnesses; OSHA 102—Summary, Occupational Injuries and Illnesses. Two of the forms, OSHA 100 and OSHA 102, are mandatory.

"First Report of Injury" forms completed for workmen's compensation purposes may in some cases be used in place of OSHA 101 provided that information required by OSHA-101 which is not contained in a "First Report of Injury" is added. OSHA 100 is a continuous log of all recordable injuries. Instructions are contained on the reverse side of the form concerning what is to be recorded.

OSHA 102 is a calendar year summary of data contained in the "Log of Occupational Injuries and Illnesses" which must be completed by January 31st for the preceding year. One copy should be retained for file purposes. One copy must be posted where employees can examine it. No length of time for posting this form has been established. It is suggested that it be placed in a glass-covered frame and remain posted for 30 days.

Each accident or health hazard that results in one or more fatalities or hospitalization of five or more employees, must be reported to the area director of the Occupational Safety and Health Administration within 48 hours.

Each recordable occupational injury or illness must be entered on OSHA 100 within two working days of receiving information that the case has occurred. Samples of each of these three forms are to be attached. All forms are to be kept at the location where employees work for a period of five years.

Forms required by OSHA should be kept in one file or folder and be available for review by a Department of Labor compliance officer. Within the publication "Recordkeeping Requirements under the Williams-Stieger Occupational Safety and Health Act of 1970" is a centerfold poster which must be displayed in a prominent place. Employers have been cited for not having this poster displayed, so it is important that it be done.

NOISE ENFORCEMENT DIVISION, U.S. ENVIRONMENTAL PROTECTION AGENCY

On April 30, 1976, the Administrator of the U.S. Environmental Protection Agency established the Noise Enforcement Division under the Deputy Assistant Administrator for Mobile Source and Noise Enforcement, Office of Enforcement. The new division currently has a staff of 21 persons whose responsibilities are divided into the following four general enforcement areas: 1. general products noise regulations; 2. surface transportation noise regulations; 3. noise enforcement testing; and 4. regional (EPA), state and local assistance.

The general products group will develop and help implement strategies and regulations for enforcement of noise emission standards for a variety of new products established pursuant to Section 6 of the Federal Noise Control Act of 1972 (NCA), PL 92-574; 86 Stat. 1234. On December 31, 1975, EPA promulgated noise standards and regulations for the control of noise from portable air compressors. These regulations will become effective on January 1, 1978. Additional regulations are currently being developed to control noise from truck-mounted solid waste compactors, truck transport refrigeration units, and wheel and track loaders and dozers, which have been identified pursuant to Section 5 of the NCA, as major noise sources.

The surface transportation group will have similar responsibilities with respect to transportation-related products. The first such products to be regulated under Section 6 for the control of noise are new medium- and heavy-duty trucks (in excess of 10,000 lb GVWR). Regulations for trucks were promulgated on April 13, 1976 and will become effective on January 1, 1978. Motorcycles and buses are additional major noise sources which have been identified and for which regulations are presently being developed.

The primary federal enforcement strategies which will be applied to Section 6 new products to assure their conformance with noise standards are the following:

1. production verification of configurations or classes of products, *i.e.*, testing of initial representative products at the assembly line
2. selective enforcement auditing of products, *i.e.*, statistical sampling and testing of new products at the assembly line
3. manufacturer's time-of-sale warranty
4. tampering prohibitions
5. maintenance instruction requirements
6. recall of noncomplying products.

Noise enforcement testing will be conducted by the new EPA Noise Enforcement Facility located in Sandusky, Ohio. This facility will be used to conduct enforcement testing, to monitor and correlate a manufacturer's compliance testing, and to train regional, state and local personnel for noise enforcement.

The regional, state and local assistance program will define and develop the EPA enforcement responsibilities under the NCA. These responsibilities will be divided between the EPA Office of Enforcement in Washington, D.C. and the 10 EPA regional offices. This program will also provide assistance to state and local agencies regarding enforcement of the federal noise control standards and regulations, and enforcement aspects of additional state and local noise control regulations.

With the primary exception of aircraft, any state or local jurisdiction may adopt and enforce time-of-sale and in-use noise regulations for new products which have not been regulated by EPA. With respect to new products which have been regulated by EPA, *e.g.*, medium- and heavy-duty trucks and portable air compressors, state and local governments may adopt and enforce time-of-sale noise emission regulations *only* if the state or local regulation is identical to the EPA regulations. However, with the exceptions of aircraft (Section 7), interstate rail carriers (Section 17) and interstate motor carriers (Section 18), any state or local jurisdiction may establish and enforce *in-use* controls on environmental noise from products through the licensing, regulation or restriction of the use, operation or movement thereof.

State and local jurisdictions may adopt and enforce in-use standards applicable to noise emissions resulting from the operation of interstate rail carriers and interstate motor carriers only if such standards are identical to the federal standards. However, a waiver of this federal preemption may be granted if the Administrator of EPA determines that such state or local regulation is necessitated by special local conditions and is not in conflict with federal regulations for interstate rail and motor carriers. Types of state and local noise regulations include the following:

1. product performance standards implemented through licensing or certification procedures
2. operational limitations, such as curfews
3. movement limitations, such as restrictions on truck traffic in noise-sensitive areas, *e.g.,* hospital zones
4. property line limitations according to land use, *e.g.,* maximum noise emission levels at property lines of residential zones
5. nuisance prohibitions.

To assist state and local governments in drafting noise control ordinances, EPA has published a Model Community Noise Ordinance. Copies are available from EPA regional offices and from the EPA headquarters in Washington, D.C.

The Noise Enforcement Division is assisting in the development and refinement of EPA policy related to NCA legal issues, including the following: 1) federal preemption pursuant to Sections 6, 8, 17 and 18; 2) special local condition waivers pursuant to Sections 17 and 18; and 3) compliance with the NCA by federal agencies.

CHAPTER 4

ACOUSTICS AND THE SOUND FIELD*

THE NATURE AND PROPAGATION OF SOUND

Sound waves are pressure variations produced as a result of a mechanical disturbance in a material medium. They can be classified into two main categories according to whether or not that medium is strained beyond its elastic limit. Our attention in these chapters will be confined to common sounds and vibrations where the medium remains within its normal elastic limits. Shock waves and other impulsive sounds which exceed these physical limits require a different approach with regard both to theory and to instrumentation.

In acoustical measurements we generally measure the pressure variations about the local ambient and static pressure of the medium, although it is possible also to measure either the local particle velocity or even particle displacement. It has been found that pressure-sensitive transducers give the most accurate electrical reproduction of the sound waveform.

The production, propagation and detection of sound waves are generally related to the setting up of an oscillation. The simplest form of oscillation is Simple Harmonic Motion; when pressure variations of this form are produced, the resulting sound is referred to as a pure tone because the pressure variations occur at only one frequency. Sound-wave propagation can be considered in the form of plane or spherical waves. Where the source is effectively a single point, waves set up are spherical. But at large distances from such a source, where the curvature of the wavefront is small, plane waves can be considered to exist. The plane wave effect is similar to what would exist if a large plane surface were to vibrate so that all points on the surface were in phase.

---

*By Anthony J. Schneider, B & K Instruments, Inc., Cleveland, Ohio.

If we now consider the simple harmonic vibration of such a plane surface in air or any other medium, as in Figure 4.1, it is apparent that movement of the surface causes a slight variation in the local pressure close to the surface. This will be transmitted through the medium at a

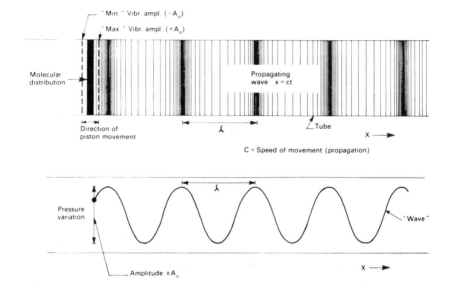

Figure 4.1 Illustration of how sinusoidal vibrations of a plane surface are transformed into a sinusoidal pressure wave.

particular velocity of propagation, as are the ripples on the surface of still water. The velocity is proportional to the elasticity and density of the medium. In an ideal gas, the relationship is:

$$V = \sqrt{\frac{\gamma \varphi}{\rho}}$$

where V is the velocity of propagation, φ the local atmospheric pressure, ρ the ratio of the principal specific heats of the gas, and γ the local density. From the gas laws, the velocity of propagation in a particular gas will be proportional to temperature according to the relation:

$$V = \sqrt{\gamma R T}$$

where R is the gas constant and T the absolute temperature. Although we normally think of sound propagation in gases, particularly in air, the basic acoustical principles apply equally well in solids. Table 4.1 gives the velocity of sound propagation in various materials.

ACOUSTICS AND THE SOUND FIELD

Table 4.1 Velocity of Sound Propagation in Various Media

Medium	Velocity of Propagation (m/sec)
Air at 1 atm 0°C	332
Air at 1 atm 100°C	386
Air at 25 atm 0°C	332
Hydrogen 0°C	1269
Vulcanized rubber	54
Granite	6000
Mild steel	5050
Aluminum	5000
Lead	1200
Water 15°C	1440
Salt water 15°C	1470
Alcohol	1213

If propagation is truly plane wave, then the variation of local pressure at a distance from a source, such as that in Figure 4.1, will also be sinusoidal. The properties of the wave are illustrated in Figures 4.2 and 4.3. Figure 4.2 shows the pressure distribution at any instant in time in the sound field, whereas Figure 4.3 shows the pressure-time variation at

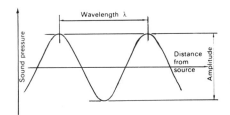

Figure 4.2 Time properties of a simple harmonic oscillation.

Figure 4.3 Distance properties of a simple harmonic oscillation.

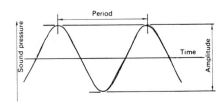

any point. While the sound pressure varies through a complete cycle in time, T, the wavefront will move with velocity of propagation V, where $V = \lambda/T$, or alternatively $V = n\lambda$ where n = frequency of the vibration, which is defined as the number of periods completed in unit time. When the period is measured in seconds the frequency is in cycles per second, or Hertz. The relationship between frequency and wavelength is shown in Figure 4.4 for the particular case of sound propagation in air.

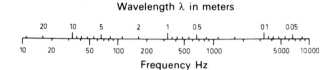

Figure 4.4 Relationship between frequency and wavelength for sound in air at normal atmospheric conditions.

The strength of the pressure wave can be measured by the amplitude of the local pressure fluctuations around the ambient atmospheric pressure. Such pressure variations are an extremely small proportion of the atmospheric pressure, and the most common unit of measurement is the Pascal (Newton per square meter) which is approximately 10^{-5} atmosphere.

THE SOUND FIELD

Although this description has been for a plane, free, progressive wave, similar concepts apply for spherical waves which are encountered in field measurements. The only difference is due to the enlarging area over which the energy of the source is spread as the wave progresses. This gives rise to the concept of intensity, which is the average acoustic power per unit area. A practical illustration of intensity is a sound which is painful when the source is close to the ear but is inaudible at a distance. Assuming that the emitted noise power is W, this spreads out from the source in the form of spheres with continuously increasing radius, r (Figure 4.5).

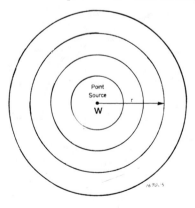

Figure 4.5 The concept of spherical wave fronts.

Thus the noise power passing through a unit area on the surface of a sphere of radius, r, must in a nondissipative medium equal the emitted power, W, divided by the area of the surface ($A = 4\pi r^2$):

$$I = \frac{W}{4\pi r^2}$$

Now, at a sufficient distance away from the source, this average noise intensity, I, is also proportional to the square of the sound pressure, p, at r, whereby:

$$p = \alpha \frac{1}{r}$$

This relationship is the inverse (distance law—sometimes also called the "inverse-square law"—and governs the free sound radiation in the acoustic "far" field) of a sound source.

Spherical radiation from sound sources and plane wave fronts at measuring points in space are ideal conditions of sound propagation which are seldom encountered in actual practice. When one or more sound-reflecting objects are present in the field, the wave picture changes completely due to the reflections. This is easily understood as there is not only the original progressive wave present, but also a reflected wave travelling more or less in the opposite direction to the original wave. The sound pressure at a certain point in the field is, at any instant, the combination of the pressure due to the original wave and the pressure(s) due to the reflected wave(s).

Before discussing the effects of reflection, the conditions necessary to obtain a reflection at all should be briefly mentioned. If for instance an object which is very small compared with the sound wavelength is placed in the sound field, no real reflection of the wave is produced. To produce a reflection which effectively interferes with the sound field, the dimensions of the reflecting object in the directions perpendicular to the direction of travel of the original sound wave must be on the order of the sound wavelength, or larger. The amount of reflection also depends upon the sound-absorbing properties of the object. Thus, both the physical dimensions and the sound absorption of an obstacle affects its reflecting properties. Returning to the effects of reflections upon the "original" sound wave one of the most important effects with regard to noise measurements is the production of a so-called diffuse sound field. A diffuse sound field is a field in which a great number of reflected waves, from all directions, combine in such a way that the average sound energy density is uniform everywhere in the field. An approximation to this kind of field exists in large, reverberant rooms, and the way in which it is achieved is demonstrated in Figure 4.6.

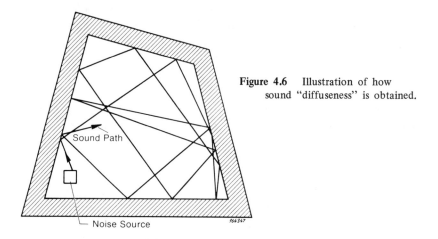

Figure 4.6 Illustration of how sound "diffuseness" is obtained.

To illustrate the importance of the above-described kinds of sound fields with respect to noise measurements, assume that a noise source is placed in a room. The sound pressure level set up around the item as a function of distance might then be of the type shown in Figure 4.7.

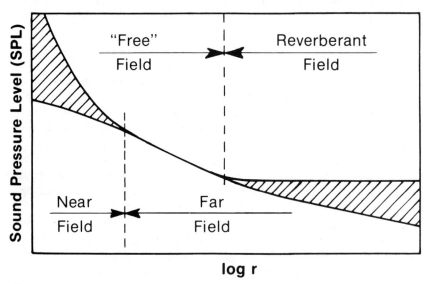

Figure 4.7 Sketch illustrating the variation of sound pressure level in a room as a function of distance from a source. The shaded areas indicate regions of level fluctuating with position.

The near and far field boundary is determined by the wave length of the sound wave and by the distance between the sound source and the point of measurement. In the near field, the sound pressure is heavily dependent upon the radiating characteristics of individual noise-generating components, and the particle velocity does not necessarily have the same direction of travel as the sound wave. The result is that small changes in position show large differences in pressure. In the far field, particle velocity is primarily in the direction of propagation of the sound wave, and pressure is inversely proportional to distance. In practice, the boundary between the near and far field is taken to be at a distance of about three times the largest dimension of the largest radiating surface of the sound source. In the far field, measurement can be made of the sound actually radiated by the source. And this data can be used to predict sound pressure at greater distances. In the near field, component sources can be measured and identified by using a probe microphone.

The free and reverberant field boundary is determined by the environment in which the acoustical measurement is made. A free field is one in which there are no reflections. This condition is often simulated in anechoic rooms for measurement of the radiated noise and the directional characteristics of sound sources. But in actual environments such as outdoors and factory areas, the ground or floor and walls reflect sound. The result is that more than one wave is present at the point of measurement. The resultant pressure is higher or lower, depending upon amplitude and phase differences between the primary wave and the reflected waves. In the free field the directly radiated sound dominates, but in the reverberant field the effect of reflections dominates. Reverberant fields are poor conditions under which to evaluate sound radiated by sound sources. But this has no bearing on environmental noise measurements made in the community or at the work place.

PHYSICAL SCALES FOR MEASUREMENTS—THE DECIBEL

In characterizing the magnitude of a sound pressure wave, the rms (root-mean-square) value is used because it has a direct relationship to the energy content of the signal. It is defined as:

$$A_{rms} = \sqrt{\frac{1}{T} \int_0^T a^2(t) dt}$$

where a is the instantaneous amplitude and T is the period of observation. The quantity normally measured when dealing with acoustic noise is therefore the rms sound pressure. Sound pressure is the force per unit area caused by the sound wave, and the unit of measurement is the Newton/m², referred to in the International Standards Organization Recommendation 1000 for S.I. units as the Pascal (Pa).

The weakest sound pressure perceived as sound is a very small value, and the range of sound pressure perceived as sound is extremely large. The weakest sound pressure disturbance to be detected by an "average" person at 1000 Hz has been found to be 20 μPa (approximately 0.003 μpsi). On the other hand, the largest sound pressure perceived without pain is of the order of 10^7 μPa, because the scale of sound pressures covers such a wide dynamic range the use of a linear pressure scale is considered to be impractical. Also, the hearing mechanism responds nonlinearly to changes in sound pressures. It is, therefore, convenient to use a nonlinear scale for sound pressure measurement. Such a scale is the decibel or dB scale.

The decibel is defined as ten times the logarithm to the base ten of the ratio between two quantities of power. As the sound power is related to the square of the sound pressure a convenient scale for sound measurements is defined as:

$$\text{Sound Pressure Level} = 10 \log (P^2/P_o^2) = 20 \log (P/P_o) \text{ dB}$$

where p is the rms level of the sound pressure being measured and p_o is a reference sound pressure, normally taken to be 20 μPa. It should be noted that the term *level* has been introduced in the above equation. This indicates that the given quantity has a certain level above a certain reference quantity. It reduces this dynamic range of 1:10,000,000 in sound pressure to an equivalent 0 to 120 dB, a much more manageable set of numbers.

In Figure 4.8 are shown some examples of the level of familiar noise sources in our environment. Circumstances often arise where it is necessary to add two or more decibel levels. Of course, it is not possible to make this addition algebraically, because of the logarithmic scale. The levels must be combined on an energy basis. The procedure for doing this is to convert the decibel values to power ratios: add or subtract these ratios as required and then reconvert to decibels. A graph which conveniently eliminates logarithmic manipulation is given in Figure 4.9. The graph shows that combining any two equal levels results in a 3-dB increase in total level. It can also be seen that if two levels differ by more than about 10 dB, then the sum of the two levels will be insignificantly different from the higher of the individual levels. From this fact is derived the rule-of-thumb that the source noise to be measured should be at least 10 dB above background level.

ACOUSTICS AND THE SOUND FIELD 49

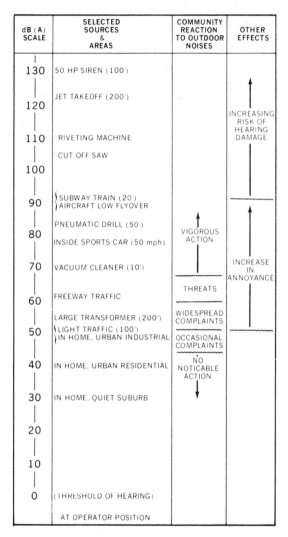

Figure 4.8 Noise levels in dBA, community reaction, and other effects for some typical sound sources.

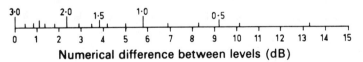

Figure 4.9 Graph for addition of two decibel levels.

LOUDNESS AND OBJECTIVE MEASUREMENTS

The human ear exhibits a nonlinear frequency response. Moreover, frequency response varies with sound pressure level. Figure 4.10 shows a set of equal loudness curves for an average young person with good hearing, listening to pure tones in a sound field. The ear is most sensitive around 3500 Hz with a marked reduction occurring at low frequencies. With increasing level, the response of the ear becomes more flat. Each curve is assigned a Phon value equal to the sound pressure intercept at 1 kHz. For example, each point on the 40 Phon curve is as loud as 40 dB SPL at 1 kHz.

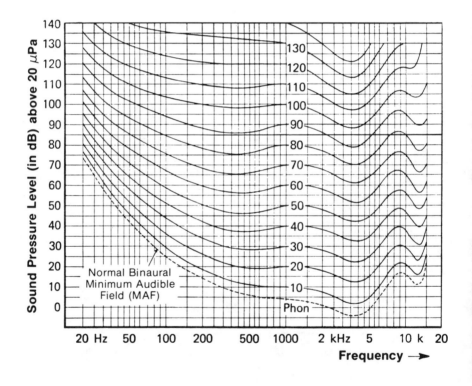

Figure 4.10 Normal equal loudness contours. They can be applied when:
    a) The source of sound is directly ahead of the listener;
    b) Listening with both ears;
    c) The sound is measured in the absence of the listener;
    d) The listeners are people with normal hearing aged between 18 and 25 years.

The ear also has a nonlinear amplitude response. An increase in Phon level of 10 dB represents an approximate doubling of perceived loudness. The frequency response of the human ear is the inverse of the equal loudness curves. Estimating loudness level on a simple instrument is done by making its frequency response have the same general shape as that of the ear's response. Sound level meters have standardized weighting filters with A, B and C response curves shown in Figure 4.11. These curves

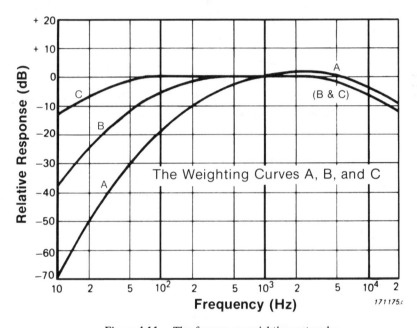

Figure 4.11 The frequency weighting networks.

simulate the response of the human ear at low, medium and high amplitudes. The original intention of the concept of the weightings was that the A scale should be used for measurement of sounds of low pressure level, while the B and then the C scale should be used at higher pressures. However, this was not completely satisfactory in practice, and in recent years the A scale has been most widely used as an indication of subjective loudness over a wide range of amplitudes and for evaluating hearing damage risk. Measurements in dBA are properly referred to as sound levels. Readings with a flat response amplifier are referred to dB SPL (sound pressure level).

COMPLEX NOISE

Sounds resulting from sinusoidal vibrations are pure tones, but the great majority of common sounds are of a more complex nature, having components at several or all frequencies in the audio range. For example, when sounds are produced from musical instruments, one fundamental frequency predominates, but others are present which are harmonically related to the fundamental and which give character to the sound of the instrument. For convenience in the analysis of such sounds with many frequency components, the sound spectrum is commonly used. This is a graph of sound pressure amplitude *vs.* frequency, averaged over a particular interval of time, in a pure tone which appears at a particular frequency as a line whose length is proportional to the amplitude of the sound pressure, as in Figure 4.12. Sounds with many harmonics, such as those of a musical instrument, or other periodic sounds appear as a number of lines at discrete frequencies (Figure 4.13). Spectra of this type are referred to

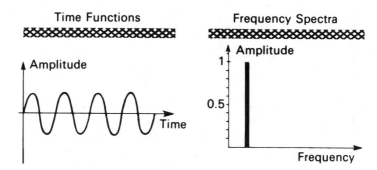

Figure 4.12 Sound spectrum of a pure tone.

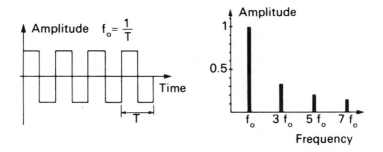

Figure 4.13 Sound spectrum of a periodic function.

as "line spectra." The majority of common sounds, however, are nonperiodic, and therefore have components at an infinite number of frequencies. These appear as a "continuous spectrum," similar to the curve of Figure 4.14. Common examples of the continuous spectrum are the hiss of an air jet, or even the background noise in the room where we sit. Frequency analysis is more thoroughly discussed in a later chapter of this book.

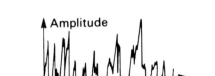
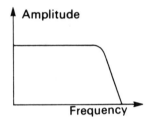

Figure 4.14 Continuous spectrum.

CHAPTER 5

ENGINEERING CONTROLS AND SYSTEMS DESIGN

Legal requirements of OSHA make it imperative that steps be taken to reduce noise. When employees are subjected to sound which exceeds the permissible exposures set forth by OSHA, feasible administrative or engineering controls must be utilized. The Department of Labor considers "feasible" to mean: capable of being done, accomplished or carried out; capable of being dealt with successfully.

Engineering controls are those which reduce the sound intensity either at the source of the noise or in the hearing zone of the workers. The ideal way of controlling noise is at the source, to prevent noise-induced hearing loss. To accomplish this, the source of the noise must be known and also its transmission path.

There are many interacting elements which make analyses of sound sources and transmission paths in industrial plants complex problems. These elements include various shapes and sizes of enclosures and equipment, the presence of diversified vibrating, radiating and reverberating surfaces, and differences in environments.

The following are engineering control measures which can be used to curtail noise: maintenance; substitution of machines; substitution of processes; vibration damping; inclusion of noise level specifications when ordering new equipment; isolating noise sources; and isolating the operator. Table 5.1 lists some typical industrial equipment noise sources.

MAINTENANCE

A method which may be obvious but many times overlooked is regular and thorough maintenance. This is essential in order to attain efficient noise control. A machine can be the best designed and built on the market, but vibration and wear may eventually cause it to become noisy. A

Table 5.1 Industrial Equipment Noise Sources

System	Source
Heaters	Combustion at burners Inspiration of premix air at burners Draft fans Ducts
Motors	Cooling air fan Cooling system Mechanical and electrical parts
Air-Fan Coolers	Fan Speed alternator Fan shroud
Centrifugal Compressors	Discharge piping and expansion joints Antisurge bypass system Intake piping and suction drum Air intake and air discharge
Screw Compressors (axial)	Intake and discharge piping Compressor and gear casings
Speed Changers	Gear meshing
Engines	Exhaust Air intake Cooling fan
Condensing tubing	Expansion joint on steam discharge line
Atmospheric Vents, Exhaust and Intakes	Discharge jet Upstream valves Compressors
Piping	Eductors Excess velocities Valves
Pumps	**Cavitation** of fluid Loose joints Piping vibration Sizing
Fans	Turbulent air flow interaction with the blades and exchanger surfaces Vortex shredding of the blades

good maintenance program should include adequate lubrication of moving parts, proper alignment, and proper balancing of all rotating members. All worn bearings, gears, clutches, and other components should be adjusted or replaced. This not only helps to cut down noise but also increases efficiency. Panel guards, mounts, and all other members which are somehow attached to a machine should be securely tightened to keep the vibration at a minimum. Fasteners should be checked periodically to make certain that they have not loosened. A lock-washer helps prevent against loosening and should be used wherever applicable.

Maintenance is an important part of engineering controls because it keeps machines running at maximum efficiency and noise at a minimum. Simple maintenance operations such as sharpening cutting tools can eliminate some noises as well as decrease cutting time. Cracks of any nature, whether in cutting tools or in machines, can produce noise and should be welded or the part replaced.

Faulty installation and/or maintenance can result in excessive vibration. Equipment should be periodically checked. Gradual increases in vibration should be examined in routine maintenance; sudden increases call for action. Increased vibration in machinery can be caused by the following:
- rotational imbalance which requires rebalancing
- misalignment of couplings or bearings
- eccentric journals
- defective or damaged gears
- bent shafts
- mechanical looseness
- faulty drive belts
- rubbing parts and resonant conditions

Rapid increases in vibration can be traced to a variety of causes, such as lack of lubrication, overload or misalignment. In pumps, cavitation may erode an impeller. In fans, dirt may adhere to the blades and then break off unevenly.

SUBSTITUTION OF MACHINES

Machines are almost constantly running at their maximum speed. This sometimes produces sounds at very high frequencies which can be eliminated with larger, slower running equipment. The rate of production can be the same and even greater, if necessary, than that of the smaller machines. With larger machines, the noise is cut down; they also require less maintenance because they do not always operate at peak capacity. Presses can be substituted for hammers, immediately reducing the amount of exposure decibels. However, such substitution can be costly and may not always be economically feasible.

Considerably quieter hydraulic and/or pneumatic equipment can be used instead of mechanical equipment. Hydraulic power supply units can be placed on antivibration mounts. Also by cutting down on piping and number of fittings required and by using flexible hoses, noise can be reduced. Some manufacturers of hydraulic equipment are promoting use of an electric motor which operates at 1200 rpm rather than 1800 rpm. A larger pump operating at lower speeds is quieter and yet delivers the same flow as a smaller unit running at a greater speed. Hydraulic power units can also be placed outside a plant or in soundproof enclosures, thereby reducing noise to a bare minimum.

Machines, such as compressors, have had major noise problems with their exhausts. This type of noise can be reduced with mufflers or silencers. Many types of mufflers are used commercially and are discussed in a later chapter. The dissipative type uses sound-absorbing material; the reactive type has various geometrical shapes to provide impedance mismatch for the acoustic energy. Some mufflers combine both effects, and performance varies with frequency of the sound and the back pressure created by the muffler.

Exhausts of air valves, cylinders, and other pneumatic devices have been equipped with silencers and are capable of reducing noise by 10 to 13 dB when operating under pressures of 80 to 100 psi. These units have hemispherical interference chambers made from sintered powdered bronze that permit contaminants and oil to pass through with air, while producing a minimum of back pressure.

Using belt drives instead of gears, noise reduction can be significant. The belt is one solid piece of material and makes relatively little noise. It runs between two pulleys under tension, whereas the gear drives using link chains are usually noisier. Gears have teeth in which the chain links must ride. The chain must be kept in tension and if tension slacks off, chain drives may jump a tooth and create a loud noise. The chain alone, just by rotating, creates more noise than the belt drive; therefore, the belt drive can be a substitute for gears.

SUBSTITUTION OF PROCESSES

Substitution of a quieter process, machine or tool is another method of controlling noise. Some examples of this method are drilling for punching, pressing or rolling for forging, hot forming for cold forming, grinding of castings instead of chipping, hydraulic presses instead of mechanical. Sometimes by substituting for a process automation may be increased; increased automation reduces the number of employees to be exposed to noise.

Welding is often mentioned as a quiet substitute for riveting, and is also faster than riveting for fastening two parts together. In riveting, sometimes the holes do not line up and then have to be redrilled. This is time-consuming and induces added noise which could have been eliminated by welding. Welding is also longer lasting. Rivets, when exposed for long time periods, can corrode or rust, leading to loss of strength and eventually failure.

There are newer methods and machines available which are orbital, spinning or radial cycloidal peening instead of impacting to form rivet heads or flares. By using these newer methods, riveting can be performed at reduced noises. A worker who is exposed to riveting all day is exposed to one of the highest noise levels on the dB scale. It is sometimes impossible to control noise at the source or path, and personal hearing protection is not acceptable as a long-term solution.

VIBRATION DAMPING

It is usually not practical or possible to isolate resonant vibrations from some metal structures such as thin panels, conveyors and chutes. Instead, they are often given a damping treatment. Damping is a restraining force opposing or reducing the amplitude of a vibration into thermal energy. Damping, however, is of little value for other than resonant vibrations. Smaller panels vibrate less than the larger ones; by providing air openings in large panels, radiated noise can be decreased.

Vibration **isolator** mats, pads or sheets can be placed between machines and foundations or floors. These isolators can be single or multiple layers, depending on their use. These isolators are made of natural or synthetic rubbers, rubberlike plastics, cork, felt, sisal, metallic wool, dense glass fibers, and other materials. Pads made of a sheet of steel sandwiched between two layers of asbestos within a sealed envelope of sheet lead are effective for high-tonnage presses and other impact machines (see Chapter 12). While pads are the simplest mountings, they may not be desirable because of restricted height adjustment, uneven contact with rough floors, or difficulties in protecting them from oil. Ribbed, waffled or studded constructions are sometimes used for protection from oil, and bolts or wedges can be incorporated in the isolators for height adjustments.

Rubber isolators may be preferable to other types of isolators because of economy, elimination of necessity for adjustment, ease of installation and replacement, and characteristics of the material. A rubber isolator's most valuable properties are its elasticity, which makes it a good shock absorber, and its ability to reduce noise transmission. Rubber has the characteristic of relatively large deformations due to loads.

Steel springs are also widely used, but generally for low-frequency isolation problems. With most common installations, springs are used to provide a natural frequency of the support mounting lower than that of the machine. In these cases the machine will pass through resonance with the support of startup or shutdown. For this reason, it is necessary to obtain a damping effect by using springs with other materials to limit movement as the machine passes through the resonance frequency.

Adjustable snubbing can also be provided to limit movement, and is especially useful in limiting movement from unusual loading.

The success of noise control is based on reduction at the source, the transmission path, or protecting the receiver. Table 5.2 shows several common means of noise abatement. Each noise problem must be handled individually. There are no hard and set rules. A problem may entail several malfunctions in a piece of equipment, and there may be more than one source. Probably the chief culprit in noise generation, and most important in transmission, is vibration. This can be the most detrimental of noise generators since it can cause machines to deteriorate faster, produce higher maintenance costs, and lead to production rejects.

Table 5.2 Control Methods for Handling Excessive Noise

Plenums	These devices admit low-velocity cooling and combustion air to noise sources. At the same time they prevent the escape of excessive noise.
Enclosures	Consist of a hard outer shell and adsorbent liner for trapping noise.
Lagging	Lining that absorbs radiated noise. It is directly applied to piping, vessels, valves and other equipment.
Damping	Reduces the amount of noise generated by vibrating surfaces.
Silencers or mutes	Attenuate noise from high-velocity flows of gases. This is done by multiple reflection of sound waves from acoustically absorbent surfaces, elimination of turbulent flow, reduction of flow velocity and cancellation of sound waves by flow interaction without employing absorption.
Ear plugs and muffs	These serve only to protect workers in the immediate vicinity of the source. They should not be considered a permanent protection, only as an expedient.

CAUSES OF EXCESSIVE VIBRATION

Unbalance is the most common cause of vibration. Loose mountings or holding devices may be a source of trouble. Differences in dimensions of mating parts, such as a shaft with a diameter that may only be minutely smaller than the corresponding bore of the rotor to be mounted, is another possible cause. Others include bent shafts, which locate the off-center from the axis of rotation, and machine errors where less material is removed from one side of a roll than from another.

Assembly errors constitute another common cause of vibration. They can be arrested by insuring surface-to-surface contact of mating parts. On fans and blowers, rebalancing should be done after worn-out fan blades have been replaced.

Misalignment can cause excessive vibration. Flexible couplings should be aligned as carefully as solid couplings. Although misaligned flexible couplings transmit torque, vibration can be so high as to cause excessive wear of bearings and journals. Proper lubrication of flexible couplings will usually correct this.

Drive belts are another source of vibration. The length and tension on a belt can generate resonance within the drive belt, causing it to behave like a vibrating string. A high axial vibration and excessive thrust bearing wear are caused by unequal tensions in multiple V-belts. The problem can be handled by replacing the belts with ones of equal tension. Proper alignment of belts will also help.

Gear problems can stem from unbalance, resonance, tooth wearing, and bearing instability. Usually, lubrication has no effect on the noise generation of a gear system. The type of oil and its viscosity are of small importance as long as contacting surfaces are coated and a dry running situation does not exist. The use of lightweight viscoelastic materials may help, but this is primarily a problem not easily handled.

Antifriction bearings contribute to noise by the out-of-round shape of the bore or shaft involving the bearing-pressed fit. This may be alleviated by utilizing a slip fit between mating surfaces. Adhesive is used to keep the bearing race from turning in the matching bore.

VIBRATION ISOLATION AND DAMPING

Reduction of vibrational forces can be handled through isolation (a separate maintenance procedure), general maintenance and modification of existing systems, and damping. General maintenance and modification procedures have been previously discussed. Maintenance modification methods can reduce vibration and sound transmission by

- replacement of worn parts
- proper re-alignment and balancing of the rotating equipment
- proper and periodic lubrication to reduce frictional forces
- tightening of parts
- proper assembly of equipment
- reduction of operating speed of equipment
- reduction of flow velocities of gases and liquids which will decrease turbulence and speed
- periodic cleaning of equipment

Isolation should be considered a key step for noise abatement and machinery operation performance. The isolation of vibration in some cases requires virtually a complete solution to the problem of noise transmission. This is particularly evident in cases of transmitted noise to the floors below the machinery or to rooms adjoining a mechanical operation. If the system is excited by a continuous vibration at one of its natural frequencies, a condition of resonance results. Under such conditions, large amplitudes of vibration are induced by very small amounts of energy. The vibration transmitted to a floor on which a particular machine is mounted may set up sympathetic movement of other machinery installed elsewhere on the floor. Decoupling of machinery from the floor with the use of vibration isolators will alleviate many problems associated with vibration and noise transmission through structures. The use of vibration isolators will also alleviate structure-borne noise problems associated with vibration. Proper selection will ensure the desired degree of isolation efficiency. Manufacturers of vibration isolators can assist in selecting proper isolators which range from cork, the original vibration and noise isolation material, to more sophisticated air-mount systems which have lower natural frequencies and thus higher isolation frequencies.

Proper application of damping materials such as plates, shells and webs, will reduce resonances that are generated in their structures. In selecting the proper damping materials, one must examine the type of structure to be damped, the frequency of vibration, operating temperature, function of the vibrating part, size of equipment, the sound level generated, and the size of the area exposed to noise and vibration.

Some of the more common materials include lead, sand, impregnated felt and loaded asphalts. These can be obtained in sheet or mosaic form. Good damping materials should display:

1. Stiffness that is compatible with the stiffness of the structure being damped. The purpose of this is so that the maximum amount of energy is extracted during bending motion.
2. Temperature resistance. Careful consideration as to operating temperatures must be given.
3. A thickness at least equal to the section of the material being damped. Sometimes several times the thickness is needed in order to obtain maximum damping.

INCLUSION OF NOISE LEVEL SPECIFICATIONS WHEN ORDERING NEW EQUIPMENT

Quieter machines are generally more expensive—on the average, their estimated cost is about 10% higher. However, it is almost always more

ENGINEERING CONTROLS AND SYSTEMS DESIGN 63

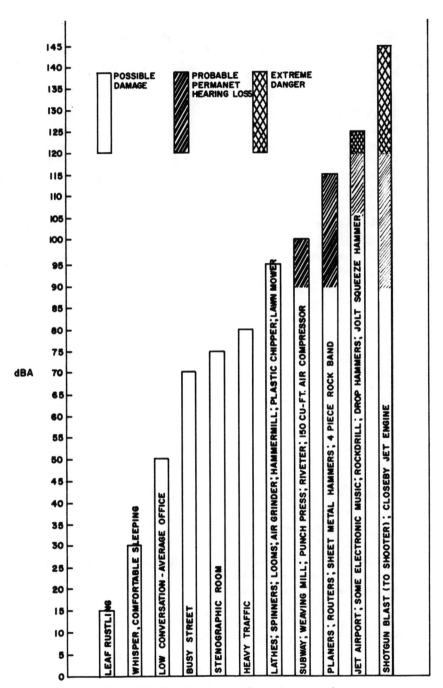

Figure 5.1 Noise level ranges for various operations.

economical to buy quieter machines than to try to reduce noise by modifying after purchase. Purchase orders should specify maximum permissible noise levels for the equipment. Major problems being encountered concern economical procurement of equipment that generates less sound and still performs the desired functions efficiently.

OSHA guidelines discussed elsewhere in this book make it almost mandatory to include noise level specifications when ordering new equipment. In the long run, money is actually saved by buying the quieter machines. Everyone profits from the quieter machines. The employees' hearing is better protected, work is performed more efficiently thus increasing production, and the employer gains from the increased production.

Since the sound produced by a machine generally increases with its horsepower, the lowest power needed for a specific application should be used. Higher-speed machines are also usually noisier, so the lowest speed required is recommended. However, improving dynamic balance can reduce noise levels at higher speeds. Operating speeds should be kept as far as possible from the resonant speeds of the machine. Acceleration and deceleration forces, as well as the mass of moving parts, should be kept to a minimum. Machines which are built with noise control specifications usually have tighter tolerances.

ISOLATING NOISE SOURCES

Controlling noise at the source is the ideal way of preventing induced hearing loss. The results are long lasting, the individual operator and his co-workers are protected, and the need for protective equipment is eliminated. A method of controlling noise at the source is to build an enclosure around the noise source. Sources can range from small pumps to very large machines, and in considering an enclosure the key to control is proper design. Factors such as environment, temperatures, durability, accessibility, personnel contact, appearance, weight and portability are all important. If an enclosure is built around an entire machine, the machine must have easy access in case of emergency. Also, the machine must be periodically lubricated, in which case accessibility is also important. An enclosure around a machine can be built with soundproof panels or soundproof doors. Also provided in the enclosure can be glass panels or doors through which vital moving parts of a machine can be observed as well as gauges and other instrumentation.

Enclosures for equipment must be constructed of proper sound-damping materials. Lead is probably one of the oldest and most common sound attenuation materials in use today, and is still popular for many applications. In general, it has two basic areas of application in sound control—

as a barrier used to block sound paths, and as a damping material for structures and surfaces to reduce transmission of vibration. Within each of these areas, there are many variations and specialized uses of lead and mixtures of lead and other materials. Perhaps the primary reasons for its continuing popularity is its extreme versatility and simplicity—even for the uninitiated. It can be cut with a common scissor, molded and shaped to any contour, and is unaffected by solvents, coolants, cutting oils and other materials commonly used in industry. It can readily be laminated to other materials such as wood, plastics and steel, and is often mixed with other materials such as epoxy and plastic to produce an effective sound absorbent with a broad range of properties. For example, lead powder mixed with various plastics and epoxies forms a compound that can be troweled on resonating machinery guards and shrouds to form a thick sound absorbent that retains flexibility and resiliency. It can be sprayed, bonded or coated onto or into metal parts and machinery to reduce and contain sound at its source.

Lead powder in vinyl with a fabric backing is commonly used as a curtain or shroud to enclose noisy operations and processes. Used as an isolation pad under equipment and even structural components in buildings, lead is unsurpassed as a noise insulator. However, lead as a sound control is perhaps most widely respected in building construction applications.

A glass barrier also offers the potential for controlling noise while providing visual communications. For the same thickness, it is a better sound barrier than most brick, tile or plaster. An average noise insulation of 35 dB can be obtained by using a single-leaf wall weighing about six pounds per square foot. However, single-leaf wall construction is not efficient if an average insulation of 40 dB or more is required. In this case a double-leaf type of wall construction must be used.

Isolating noise sources is one of the best methods for controlling noise. This technique should be used wherever possible and feasible. But there are many instances where the noise source cannot be isolated. In this case, noise should be controlled in its path or, as a final step, the operator can be isolated.

ISOLATING THE OPERATOR

As a solution to noise control in some cases, the operator must be isolated. There are some areas in which noise levels are much too high and cause severe hearing loss in a short period of time. The operator might then be isolated in a soundproof booth. The booth can be made as large or as small as necessary. If the operation is simple, a small booth will do the job. All the controls required by the operator can be placed inside the booth.

When the noise level within a confined area is too high to allow workers into the area even with the use of personal protection devices, the operation might be automated. An automated process can be supervised from observation posts, that is, from remote-control stations where the workers are adequately protected. A remote-control post can receive information via closed-circuit television, or it can be a highly insulated area within the department. The production procedures are handled by mechanical devices under man or computer control. Rolling mills are controlled from completely soundproof cabins. The same is true of workshops used for assembling and testing engines. The noise of such operations normally cannot be stifled at the source.

Engineering controls are effective in noise control and should be used wherever applicable. The correct engineering control should be used to obtain efficient results. By reducing noises, the operation is usually increased and machines run much more efficiently. Workers actually work at greater productivity with less noise.

OSHA is forcing employers to use engineering controls. New plants are being designed with noise control in mind which will effect savings as well as implement safety in the long run. Machines are also being designed to run more quietly. Engineering control and future engineering technology will help to keep noise at a minimum.

DESIGN OF SYSTEMS

In designing a system for reducing noise engineering, controls should be used wherever applicable. These controls can be effective if used properly. Systems, however, are much more complicated and must be given much more consideration. Instead of just using engineering controls to quiet noise, the placing of the machines is also a vital factor. If machines are layed out too closely together, the operator may be exposed to an excessively high dB level, whereas if spaced adequately apart, noise levels will be within acceptable limits.

In designing a system for office areas careful attention should be given to the layout design, creating an environment that will attract and keep employees. Sound control for ceilings in offices must be planned at the architectural stage. It becomes very expensive and time-consuming to soundproof an office after it is built.

Wall and ceiling construction should provide approximately the same degree of sound control through each assembly. Where a drywall partition is used for sound isolation, the construction should extend from slab to slab. Sound can travel through a suspended acoustical ceiling, up and over a partition attached to a suspended ceiling and into other areas.

Optimum sound isolation requires that the integrity of drywall partitions and ceilings not be violated by cutting holes for vents or grills or by recessing cabinets, light fixtures, etc. Instead, they should be surface mounted. Door and borrowed light openings are not recommended in party walls. The septum layer in the triple solid drywall partition should always remain whole.

Where holes are necessary, avoid placing them back to back and immediately next to each other. Electrical boxes should be staggered, preferably at least one stud space. A nonhardening, nonskinning resilient caulking material should be used to seal all cutouts, such as around electrical and telephone outlets, plumbing escutcheons, and wall cabinets. The backs and sides of electrical boxes should also be caulked to prevent sound leaks. Caulking should be used to seal all intersections with the adjoining structure, such as under-floor and ceiling runner tracks, around the perimeter at which the assembly meets the floor, ceiling, partition or exterior wall at a vertical intersection occurring at columns and window mullions.

Avoid construction such as ducts, rigid conduits or corridors, which act as speaking tubes to transmit sound from one area to another. Common supply and return ducts should have sound attenuation lines. Conduits should be sealed. Doors leading to a common hall should be gasketed around the perimeter and should not contain return air grills.

To isolate structure-borne vibrations and sound, resilient ceiling systems and floor coverings are recommended. Vibrating or noisy equipment should have resilient mountings to minimize sound transfer to structural materials. Ducts, pipes and conduits should be broken with resilient, nonrigid boots or flexible couplings where they leave vibrating equipment; they should be isolated from the structure with resilient gasketing and caulking where they pass through walls, floors or other building surfaces.

Since production area ceilings are high, office walls are often constructed of bricks or acoustical partitions. The ceiling is then suspended from the underside of the plant roof. The office walls will usually prove to be good sound barriers. However, the sound-deadening capacity of the ceiling will be far below what is required for conversation, even when acoustical ceiling tile is used in the T-bar suspension systems. The ceiling will control sound originating with the office, but will not serve as a barrier to external noise.

Noise from production equipment can reflect off the underside of the plant roof and focus on the acoustical tile ceiling of the office. As a result of this unique focusing problem, noise levels within the office can be more distracting than in the plant.

Constructing full-height insulated office walls to the underside of the plant roof is usually uneconomical in high-ceiling areas. High cost of materials and labor, interruption to production, and modifications to fire-control sprinkler systems usually make high-wall office systems prohibitive in cost.

Using a lead powder-loaded, vinyl-coated fabric can help reduce outside noise substantially. The material is limp, dense and nonporous and is easily cut with a knife or scissors. It weighs about 0.9 pounds per square foot and is approximately 1/16-inch thick. Most important, the material can be installed easily over one office in an entire group, and its effectiveness evaluated before making a commitment for a total installation.

The lead-loaded fabric can be installed by plant personnel, quickly and easily, usually with only oral instructions. Rolls of vinyl are unrolled on the floor, cut to length, and laid directly over the T-bar suspension of the ceiling. No additional support is necessary. Noise levels ranging from 95-100 dB in offices have been reduced after installation of lead-loaded fabrics to 75-80 dB.

CHAPTER 6

PERSONAL SAFETY DEVICES

Often it is not possible to stop the generation of noise at its source or by the use of acoustical materials and noise barriers. In such cases, individual protection against the troublesome noise is the only alternative.

THE ACOUSTIC PROBLEM

Individual protection against noise is accomplished by the use of ear protective devices. The function of such devices is to place a barrier between the noise source and the hearing mechanism within the ear. This barrier is a physical device which is either worn over or inserted into the external ear. This barrier will reduce the amount of noise energy transmitted to the inner ear. The degree of protection available through such devices depends on a number of physical and physiological factors that control the behavior of the sound energy as it meets the protective barrier. Figure 6.1 illustrates the acoustic problem by a block diagram.

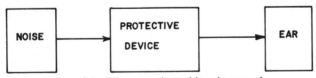

Figure 6.1   The acoustic problem in general.

Figure 6.1, however, does not fully illustrate the complexity of the factors which determine the attenuation characteristics of the barrier because sound energy reaches the inner ear by three different paths. These three paths are: through vibration of the barrier, by acoustic leaks through or around the barrier, and through bone conduction which in effect short circuits the barrier. The importance of these pathways is dependent upon the frequency and intensity of the noise and the amount of leakage around or through the protective device. Figure 6.2 illustrates a block diagram of these three paths.

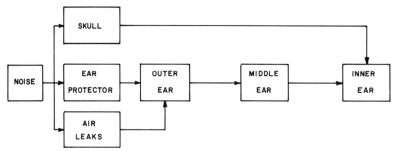

Figure 6.2 The three different paths by which sound energy can reach the inner ear, while a protective device is being used.

Where the noise airborne path is blocked completely, the noise attenuation afforded the "inner ear" is limited to approximately 50 dB below the environmental level. This attenuation limit is the result of the conduction of surrounding noises by the skull to the "inner ear." One can see how bone conduction varies with frequency in curve C of Figure 6.3. In general, a 50-dB attenuation is more than satisfactory for sufficient ear protection in an industrial environment. Therefore, the limitation due to bone conduction is neglected, except in very extreme cases. Leakage on

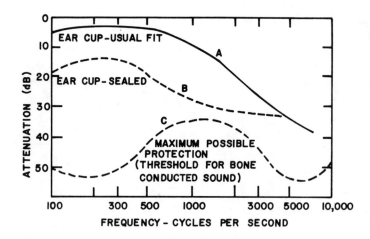

Figure 6.3 Variation of bone conduction with frequency.

the other hand, cannot be neglected and must be corrected. Curves A and B of Figure 6.3 illustrate the attenuation losses due to leakage into an improperly sealed ear muff. A poor acoustical seal greatly affects the attenuation characteristics of an insert-type ear protector (see Figure 6.4).

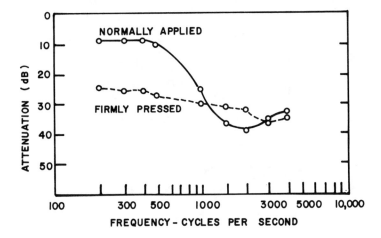

Figure 6.4 Effects of a poor acoustical seal on the attenuation quality of an insert-type ear protector.

EAR PROTECTOR REQUIREMENTS

Many surveys of industrial environments reveal that dangerous noises are generally classified in two separate categories. These two categories are: the moderate- to high-level noises of most common manufacturing areas, and the extremely noisy areas such as jet engine testing areas. A large percentage of the overall noise levels in the general manufacturing industries averages around 100 dB. In the extremely noisy industries, levels reach in excess of 130 dB.

The noises which fall into the two general categories have their own individual characteristic frequency spectra. For any particular exposure to any noise, the characteristic noise spectrum should be compared with the adopted damage risk criteria to determine if a protective device will be expected to give adequate ear protection. Figure 6.5 illustrates the noise levels which can be tolerated by a partially protected ear.

By adding the attenuation expected of the protective device to the criteria, the highest noise levels at which typical protective devices afford sufficient protection can be determined. Therefore, Figure 6.5 indicates

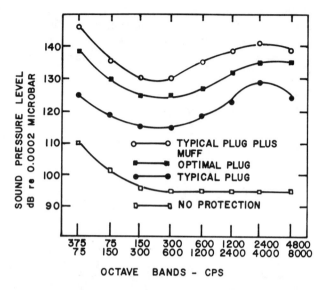

Figure 6.5 Tolerance levels of the human ear while protected in different ways.

that a relatively good ear protector will be sufficient for most of the noises in the general manufacturing industries. However, in the extremely noisy industries a combination of several protective devices should be used to ensure adequate attenuation. In addition to providing adequate attenuation, an ear protector must be nonirritating to the skin, comfortable and sanitary over long periods of time. There are two general types of ear protectors used widely by industry: the cup muff type and the plug insert type. Figure 6.6 illustrates some of these protectors.

Insert Type

Ear plugs are inner ear protection devices which are designed to occlude the ear canal. They are made from soft materials, particularly rubber, neoprene, wax, cotton, fiberglass or plastic. The choice between plugs and muffs depends greatly on the work situation and employee's preference. Both types offer adequate protection at frequencies above 1000 Hz. If an employee uses a helmet, he would probably be more comfortable with ear plugs.

There are several problems encountered by workers using ear plugs. In order to get a good acoustical seal against the sensitive inner lining of the ear canals, the inserts have to apply a certain degree of pressure. This

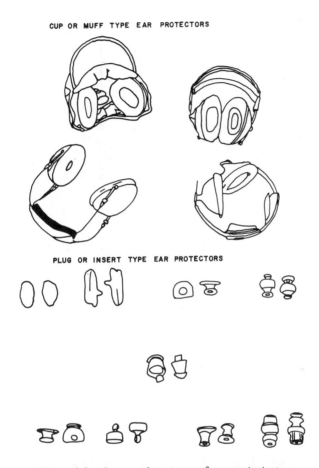

Figure 6.6 Some various types of ear protectors.

pressure can cause discomfort and is the main reason why workers find them objectionable. Many employees often prefer to use cotton as a sound suppressor; however, this is a poor alternative. Figure 6.7 illustrates the attenuation level of pure cotton. Most ear plugs provide more than double the protection of cotton. Employers should discourage employees from improvising any type of inserts.

Another problem is that ear plugs usually pick up a varying amount of dirt. Wearing ear plugs for an extended period of time may cause a "plugged" feeling, dizziness or vertigo. When such a situation arises, the worker usually places the plugs on the work area where they can pick up dirt, metal filings, germs and other debris. The only way to combat this

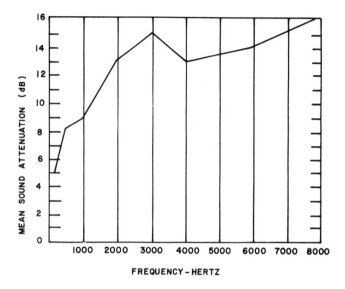

Figure 6.7 Attenuation levels of pure cotton.

problem is to fully educate the employee. It must be shown to him that it is necessary to clean the plugs every time they are to be used, unless taken directly from their container. Because of the simplicity of design, ear plugs can be readily cleaned of any dirt they pick up.

Another problem is that ear plugs might lose their efficiency during the day, due to a break in the acoustic seal between the ear and insert. This occurs because of jaw movements which tend to change the shape of the ear canal. This problem has been cured to a degree by utilizing a headband connecting the ear plugs. Lightweight headbands provide a small pressure on the canal caps to hold the plugs in place. When the band is properly adjusted to fit the head size and the plugs are properly inserted, the caps require less reseating after short periods of use.

Ear canals of different people vary in size. Most manufacturers provide standardized inserts in different sizes. However, this may cause some problems in fitting. The best method of effective control is to obtain personally molded ear plugs. There are kits sold on the market with soft putty-like material that can be molded in the individual's ear canal. This method thereby eliminates excessive pressure against the sensitive canal walls and gives maximum comfort. There are several problems, however, which can arise even when using molded types, one of which is that air can be trapped in the ear when first molding the inserts. This usually causes voids in the plugs which will prevent a good acoustical seal. It is very important that the plugs fit properly. Even the slightest leakage will lower the amount of attenuation as much as 15 dB at some frequencies. Some materials also might shrink when they harden, therefore eliminating the seal. One should evaluate all literature on such devices. Custom-made inserts should be fitted by trained technicians, nurses or doctors.

The use of personal ear protectors involves more than merely purchasing and fitting. Employees' ears should be examined and tested before fitting with ear protectors; plugs should be designed and fitted individually for each ear. Workers should be educated to the necessity of using these devices and their care. There may be a certain amount of discomfort on the part of the wearer when he first uses the ear protectors, and he should be told of this prior to issue. To obtain better acceptance employees should be given a choice of several different styles of ear protectors. One of the most important factors is to purchase the type which will be worn most effectively.

MUFF-TYPE PROTECTORS

Ear muff protectors give over-the-ear protection against dangerous high-frequency noises. However, they should also allow the wearer to hear low-frequency sounds such as voices and warning signals. This poses a problem in hearing protection equipment. Workers have to be able to communicate with each other normally, or they may not use the devices. In choosing the proper ear muffs, an engineer must determine the sound levels generated in the work area. Such data may be obtained by using sound-measuring meters. Then the attenuation levels of the ear muffs chosen must be carefully determined.

Attenuation is a measure, in **decibels**, of how effective the device is in diminishing harmful high-frequency noises. The engineer can compare hearing protection devices by comparing the attenuation data that are usually supplied by the manufacturers. Figure 6.8 illustrates a common attenuation graph. An ideal muff should be highly effective in the high-frequency ranges because it is at these frequencies that most hearing damage occurs. Attenuation provided by ear muffs varies due to differences in size, shape, seal material, shell mass and method of suspension. Another important aspect of attenuation is the wearer's head shape and size. The cushion between the shell and head also has a great deal to do with the attenuation frequency.

Many types of ear muffs and cups are available. There are basically four parts to an ear muff device. The ear cups are the outermost covering designed to deflect as much noise as possible. These should be made

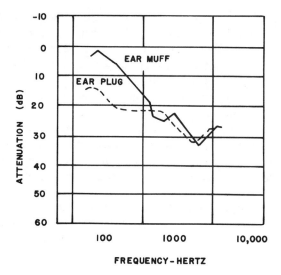

Figure 6.8 A common attenuation graph.

of a rugged material capable of withstanding any mechanical shock. Although many manufacturers indicate that ear cups are shaped to fit all ear sizes comfortably, this is not always the case. Each worker should be fitted individually. Another part is the sound-absorbent earcup insert. The outer earcups are lined with a soft foam, such as a vinyl or polyurethane foam. This lining insert serves two purposes; one, it provides additional noise reduction, and second, it eliminates the "sea shell" effect of roaring noise generated when one holds an unlined cup over the ear. A third part is the sound-tight ear cushion, which is optional on some ear muffs. This reduces sound levels and prevents sound leaks by acting similar to a washer and is usually constructed of a tough vinyl cushion. The fourth part is the headframe which connects the two ear muffs and suspends them from the head. The frame is easily adjusted with a type of slide lock mechanism. It should be light and have adequate tensile strength. One of the most important functions of a headframe is that it will provide a comfortable fit. It is necessary to provide a headframe that offers swivel action around the cups to eliminate bulkiness; in this way the wearer can adjust the headframe to provide the most comfortable fit.

Ear muffs also pose another problem; namely, they may interfere with other head gear. The swivel-action yoke headframe permits the ear cups to be worn with hard hats, bump caps and welding helmets while other models can be fastened to safety caps. A spring tension band holds the complete unit on the cap with minimal pressure; the ear cups grasp the underside of the cap brim.

The most important aspect of personal protective devices is that they depend on an effective seal between the skin and ear protector's surface. If even a small sound leak is present, the effectiveness and purpose of the ear muff is destroyed. A common disturbance encountered by users of ear muffs is that the muff might become loose as a result of talking, chewing or simple movement. There is nothing that can be done to combat this problem except to reseat the cups occasionally.

Sanitation is another aspect in using ear muffs, although this problem is more typical of inner ear protectors. It is most important that the unit be easily disassembled for cleaning and sanitizing.

Most commercial ear muffs offered on the market give about the same degree of protection. The best device is one that is accepted by the worker and is worn properly. Personal fittings should be made to ensure comfort and effectiveness. One way of determining whether ear cups are performing effectively is to conduct hearing tests on workers before and after use. When properly worn and maintained, ear muffs can provide adequate protection against most industrial noise exposure. Figures 6.9 through 6.12 show some typical industrial insert- and muff-type hearing protectors.

Figure 6.9 Insert-type ear protector.

78 INDUSTRIAL NOISE CONTROL HANDBOOK

Figure 6.10 Construction helmet with muff-type protectors fastened to the sides.

Figure 6.11 Muff-type ear protectors with flexible head strap.

Figure 6.12 Muff-type ear protectors with rigid headband.

COMBINED USE OF MUFFS AND INSERTS

If maximum protection from severe noise is required, both muff- and insert-type protectors can be worn at the same time. The attenuation given by both will not be the sum of the two devices individually, however; a little less than this value can be achieved. The difference between the upper and lower curves in Figure 6.5 indicates the noise reduction that should be expected. In no case will the total attenuation be greater than approximately 50 dB, because at this point bone conduction becomes significant.

COMMUNICATION WHILE WEARING EAR PROTECTORS

Communication is essential in any industrial environment; therefore, an important aspect of an ear protector is its ability to allow maximum communication in conjunction with adequate articulation protection. In Figure 6.13 we see the degree of speech intelligibility articulation that is expected from normal ears—both with and without ear protectors—in several background noise levels. The devices used in this investigation

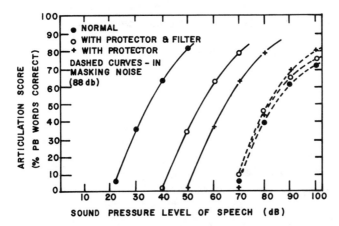

Figure 6.13 Degree of speech intelligibility with and without ear protectors.

were a soft plastic insert type which have high attenuation values throughout the audible spectrum. The curves indicate that in steady-state background noise levels, up to about 85 dB, speech intelligibility is the highest when there is no protection. In background noises above 85 dB, speech intelligibility is gained if ear protectors are worn.

Figure 6.14 illustrates the results of a study in which the intelligibility scores obtained by a subject wearing protectors, similar to those used in Kryter's study, were compared to the scores obtained with protectors having speech filters. The protectors that had speech filters were

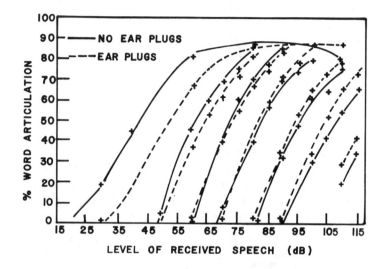

Figure 6.14 Results of a study of intelligibility scores.

constructed to give minimum attenuation of low frequencies where speech components are present. Speech intelligibility was tested in quiet surroundings and then in noisy surroundings. The results of these studies show that there is little difference in speech intelligibility scores with or without ear protectors with background noises of approximately 88 dB. In background noises below the level of 88 dB, intelligibility is highest without protectors; but in noises above 88 dB, intelligibility is better with the protectors. The protector with the speech filter indicates better articulation scores than the broad-band attenuator only in background noises below the 88 dB level.

AN EAR PROTECTION PROGRAM

To institute an effective ear protection program, the industrial hygienist must educate the management and the employees to the complete noise problem before proceeding to define the benefits of ear protectors. Management must recognize that excessive noise produces hearing loss and, if high noise levels cannot be reduced, ear protection represents an effective solution to the problem.

The worker is entitled to an explanation of the noise problem in his working environment. It is important that this explanation does not produce apathy. Employees should understand that it is unlikely that total hearing loss will result from excessive noise exposures, but partial hearing losses are quite possible, and in the early stages these losses are not readily detectable. Emphasis should be placed on the fact that a good ear protector will give sufficient protection for all but the most extreme noise exposures. The use of makeshift ear plugs such as cotton should be discouraged because they do not always provide sufficient protection. It should also be emphasized to the employee that in the typical factor environment the use of ear protectors will not deprive him of necessary hearing.

It is preferred that the distribution of ear protectors lies within the responsibility of the medical department. An examination of the ears should be carried out before the fitting of ear plugs. Any disapproval of their use should be noted and recommendations made to supply the employee with muffs or the assignment to a quieter working area.

So that the employee will get maximum benefit from an insert-type protector, it is very important that he be fitted with the correct size. After being fitted the employees should be given instructions on sanitation. Permanent ear protectors should be cleaned daily with soap and water, so that all dirt and wax deposits are cleaned off before the plugs are reinserted. A worker should periodically examine his ear protectors for any defects. If any devices are found to be unsatisfactory, they should immediately be replaced.

The medical or safety department should have available for distribution two or more satisfactory types of ear protective devices. This has a psychological advantage in that it allows the employee to choose for himself the device which gives him the most comfort and protection. The measurement of ear protector efficiency is a complicated and expensive procedure that industry cannot economically perform on every employee. The most practical method of determining the effectiveness of an ear protector is by periodic audiometric examination of the workers who use protectors.

During the first year of a hearing conservation program a number of audiometric examinations should be conducted. Afterwards, hearing tests should be conducted along with periodic physical examinations. If an audiometric examination indicates a significant threshold shift, an investigation into the cause is indicated. The investigator should determine whether the employee is regularly using the ear protector; the noise exposure has changed; the ear protectors are fitted properly; and the individual is noise-sensitive.

Supervision is a key factor to the success of an ear protection program. While the general guidance may be the task of the medical, safety or industrial hygiene personnel, supervision should be the responsibility of the plant foreman. Supervision is particularly essential in locations where the insert-type protectors are employed. Here, the foreman is the only person who is continually close enough to the employees to see that the protective devices are being used. The foreman is one of the most important men in an ear protection program.

CHAPTER 7

ENCLOSURES, SHIELDS AND BARRIERS—DESIGNING WITH LEAD

In industry today, a practical and efficient method of reducing noise in a system is to enclose it, thus cutting off some of the sound waves. Enclosures can be constructed from most common building materials. However, the degree of noise reduction is dependent upon the surface weight, as well as the internal damping and stiffness of the material.

THE USE OF LEAD AS A NOISE BARRIER

A wide selection of products and materials is available for noise reduction, one of which is lead. Lead is used to combat noise pollution in a variety of applications. The metal is applied alone and in combination with various other materials to meet OSHA regulations in numerous industrial, business, educational and residential complexes. Table 7.1 lists types of lead-containing materials that are used for industrial noise control. Lead can be specified by sheet weight (1 square foot of $1/64$th-inch sheet lead weighs 1 pound). The type of lead alloy does not have to be specified because the alloy content does not significantly affect the weight or the efficiency of lead as a noise barrier.

Acoustical properties of lead for controlling noise are very good. Qualities such as limpness, mass and internal damping are essential for decreasing and controlling sound, and lead possesses all three. Lead can be incorporated into other noise-reduction materials and, due to its mass, it can easily be pounded into thin sheets for application—an important design characteristic when maximum space is to be utilized. Mass and limpness should be considered when installing lead or lead-loaded fabric enclosures because these two qualities are most important in sound reduction. Basic considerations and procedures that should be followed for proper installation of enclosures are listed here.

Table 7.1 Lead Materials for Noise Insulation

Material	Description	Uses
Sheet lead	Usual weight $1/2$ lb to 4 lb	Used alone or laminated to substrates of various types
Lead/foam	$1/2$-lb and 1-lb sheet lead sandwiched between layers of polyurethane foam	Laminated to enclosures
Leaded plastic	Lead-loaded sheet vinyl or neoprene with or without fabric reinforcement	As a curtain or to line enclosures
Damping tile	Lead-loaded epoxy or urethane tiles	Damping heavy machinery
Casting compounds	Lead-loaded epoxy	Potting, filling complex voids
Troweling and damping compounds	Lead-loaded epoxy and urethane	Damping enclosures, surface resonant members and rattling panels

1. When installed, sheet lead should not be fastened rigidly to stiffer surface skins because of their tendency to degenerate the limp qualities of the lead.
2. When lead is laminated to another material, its weight should be kept in a composite skin nearly equal to, or in excess of, the weight of the other material.
3. Viscoelastic adhesives or intermittent fasteners should be used when laminating lead panels, rather than continuous fastenings.
4. Leaded panel skins should be used for double-wall construction instead of one skin even when the total weight of both is the same.
5. All seams, doors and perimeter joints should be caulked or fitted with gaskets to eliminate entrance of any noise.
6. Whenever sound can circumvent a lead shield, all passages should be thoroughly covered. These passages include back-to-back panels, windows, cabinets, electrical outlets, ventilation ducts, the space above a suspended ceiling, and any cracks in walls, however small. All of these passages should be completely blocked with leaded materials.

Sheet lead used for industrial noise control is specified in pounds per square foot. Manufacturers of such sheets make them available in weight from $1/2$ pound per square foot at $1/128$-inch thickness to 8 pounds at

$1/8$-inch thickness. Sheet lead is easily altered to any dimensional need; sheets can be cut with scissors, molded and contoured by hand, and applied to most surfaces with adhesives, as well as laminated to substrates such as aluminum and steel.

Another material used for industrial sound attenuation is sheet lead and polyurethane sandwich material, which can be applied in such places as the inside of machine guards or shrouds. The lead-foam material is available from the manufacturer with $1/2$-pound or 1-pound per square foot sheet lead laminated between various thicknesses of polyurethane foam. Thickness of the foam can be obtained anywhere from $1/4$ inch to 2 inches in increments of $1/4$ inch. Lead-foam material can be obtained with either two or three layers of sheet lead within the laminate. This material may be cut with ordinary scissors or steel rule dies when repetitive parts are required. The material can be anchored down with adhesives and can be shaped and molded.

Another sound control material used widely by industry is leaded vinyl sheet, an example of which is two sheets of lead-loaded vinyl laminated to a core of glass fiber cloth to give added strength and durability. This material is specified by pounds per square foot and is available in 0.5-, 0.75-, 1.0-, 1.5- and 3.0-pound weights. It is commonly applied in the same way as sheet lead and can be used with acoustical wool. Lead-loaded epoxy is another type of sound-reducing material. Powdered lead mixed with epoxies makes an excellent damping compound to control structure-borne noise. This substance can be applied by troweling it to surfaces to reduce noise and vibration.

PLENUM BARRIERS

The use of suspended ceilings is very popular in modern office construction. However, this type of ceiling installation allows a space or plenum between it and the previous ceiling from which it is hung. An acoustical problem may result, usually because the material employed in a suspended ceiling is lightweight and does not provide an adequate sound barrier.

A better system is to use plywood-lead sheets, with 2-lb lead, around a frame. These plywood-lead sheets can be covered with different types of wood panels. Office doors can also be made of the same materials. An office with lead-plywood ceiling panels and doors is more efficient in achieving sound attenuation; however, any sound made within the office will go through the ceiling and into the open space above. The noise will then find its way through the ceilings of other offices. The flow of sound through the plenum can be stopped by hanging plenum barriers. Installing such systems does not require excessive materials and is more feasible than laying sheet lead over ceiling tiles.

86 INDUSTRIAL NOISE CONTROL HANDBOOK

Plenum barriers are also used for sound attenuation in piping and air duct systems. In such systems, holes are cut into the barrier, thus leaving an air leak around the pipe or duct that should be caulked or taped. In addition, the pipes and ducts may be lined with lead to lessen the radiated noise. To develop such a noise attenuation system the pipes should first be insulated with a paper-backed fiberglass material or mineral wood insulation applied tight enough to support lead. Next, lead sheets are wrapped around the pipe in sections with each piece overlapping the preceding one by $1^1/_2$ inches. The lead sheets are sealed with a standing seam with the overlap taped. For air ducts, a resin-bonded fiberglass board is commonly employed. This material is normally available in 2 by 4-feet boards. The lead is then applied to the ducts in the same manner as for the pipes. Sheet lead plenum barriers are widely used for sound attenuation in office systems. Figure 7.1 illustrates a sound-stopping system employing plenum barriers.

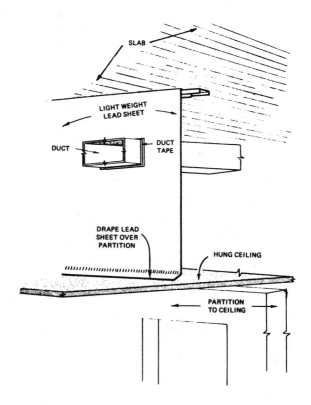

Figure 7.1 Sheet lead plenum barrier system.

ENCLOSURES, SHIELDS AND BARRIERS—DESIGNING WITH LEAD 87

HOW TO INSTALL PLENUM BARRIERS[1]

No special tools or skills are needed for installation of plenum barriers. The material is very malleable, $1/64$-inch-thick, and weighs 1 pound per square foot. Lead will stick with adhesives and can also be taped or cemented into place. Three methods of installing sheet lead plenum barriers are discussed here.

For method number one, first cut the sheet lead 4 inches longer than the total height of the plenum. Then make notches in the top corners about $1\frac{1}{2}$ x 2 inches in size. Then trim the 2-inch edge over a batten length of metal that is equivalent in length to the width of the sheet minus 3 inches. Lift the sheet with batten and nail on 6-inch centers to the blocking or the soffit of the slab. Finally, hang the bottom end of the lead sheet approximately 2 inches out over the ceiling construction. This method is illustrated in Figure 7.2.

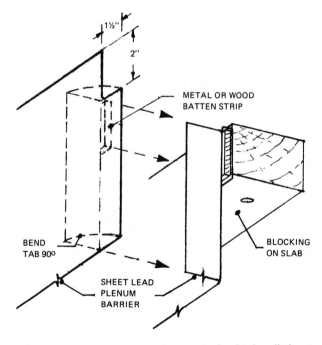

Figure 7.2 Method number one for plenum barrier installation.

In method number two, first cut the sheet lead 7 inches longer than the total height of the plenum. Make notches at the top two corners $1\frac{1}{2}$ x 5 inches in area and then turn the 5-inch edge $1\frac{1}{2}$ times around a wooden batten strip that is 3 inches shorter than the width of the sheet. Now lift the sheet with batten to the slab and fasten it on 12-inch centers with screws. Finally, hang the bottom 2 inches of the sheet over the ceiling construction. This method is illustrated in Figure 7.3.

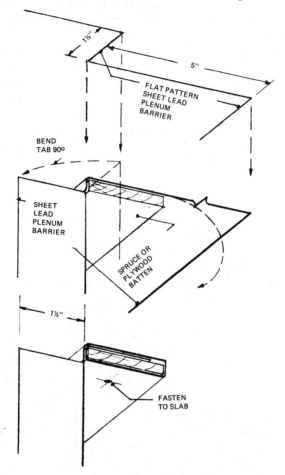

Figure 7.3 Method number two for plenum barrier installation.

In method number three, cut the lead sheet 5 inches longer than the height of the plenum. Then make $1^1/_2$- x 3-inch notches in each top corner and turn the 3-inch edge over a $1^1/_2$-inch black iron channel that is 3 inches shorter than the width of the sheet. Lift the sheet with channel to the slab and fasten down on 24-inch centers with screws. Then hand the bottom 2 inches of the sheet over the ceiling construction. This method is illustrated in Figure 7.4.

ENCLOSURES, SHIELDS AND BARRIERS—DESIGNING WITH LEAD 89

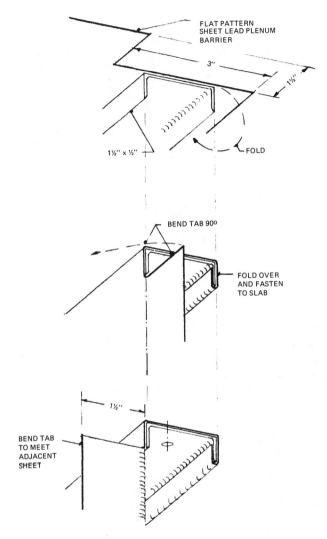

Figure 7.4 Method number three for plenum barrier installation.

When the plenum barrier is installed, seams can be formed by bending a $1^1/_2$-inch tab located at each vertical edge and sealing it with 3-inch tape. This procedure is illustrated in Figure 7.5. For the barrier to give maximum performance in sound attenuation, it must be completely airtight, which can be achieved by gasketing and caulking at all joints.

90 INDUSTRIAL NOISE CONTROL HANDBOOK

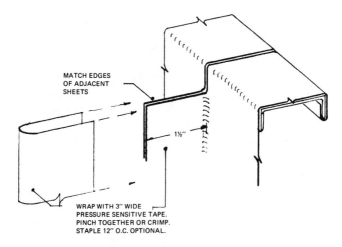

Figure 7.5 Formation of seams when installing plenum barriers.

INSTALLATION OF BARRIERS AROUND PIPES AND WIRES

When installing a plenum barrier, pipes and wires may often interfere. To circumvent such obstacles without sacrificing efficiency of the barrier, the following procedure may be employed. First, cut the lead sheet from the bottom to the point at which the pipe or wire will enter. Next, make slits in the sheet to provide for the pipe circumference. Then, peel back the segments made by the slits and slide the wire or pipe through the hole. Push the segments back against the wire or pipe so that they are snug against it. Finally, tape around the segments and the slit to the bottom of the sheet. This method is shown in Figures 7.6 through 7.9.

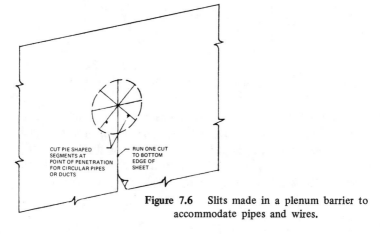

Figure 7.6 Slits made in a plenum barrier to accommodate pipes and wires.

ENCLOSURES, SHIELDS AND BARRIERS—DESIGNING WITH LEAD 91

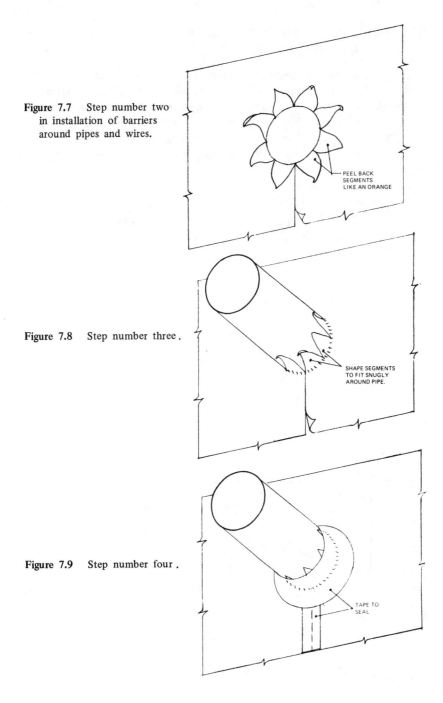

Figure 7.7 Step number two in installation of barriers around pipes and wires.

Figure 7.8 Step number three.

Figure 7.9 Step number four.

INSTALLATION OF BARRIERS AROUND RECTANGULAR DUCTWORK

To work around rectangular air ducts, the following procedure may be used. Figures 7.10 and 7.11 illustrate the procedure. First, attach adjoining sheets except for making a standing seam at the vertical edges as shown in the figures. Slit the barrier sheets as illustrated, and bend the flaps to fit the duct. Then tape the duct in place, bend back and tape the flaps. Next, form and tape a standing seam at the vertical edges.

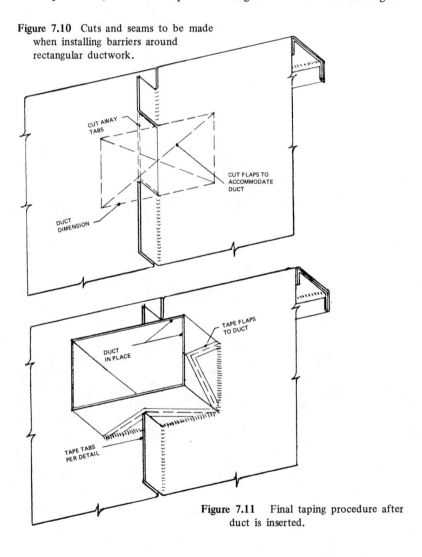

Figure 7.10 Cuts and seams to be made when installing barriers around rectangular ductwork.

Figure 7.11 Final taping procedure after duct is inserted.

ENCLOSURES, SHIELDS AND BARRIERS–DESIGNING WITH LEAD   93

BARRIERS UNDER TWO-WAY JOIST SYSTEMS

To install barriers under pan-type floors the method illustrated in Figures 7.12 and 7.13 can be employed. First, fasten the lead sheet to the slab. Then cut sections of 1-pound sheet lead to place into the cavities of the floor construction. Finally, tape the bottom edge to ground.

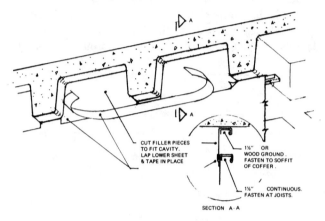

Figure 7.12   Step number one in installing plenum barriers under two-way joist systems.

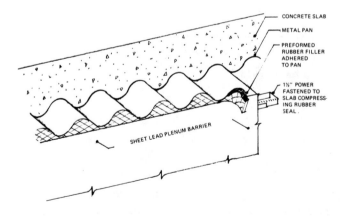

Figure 7.13   Step number two.

HIGH TRANSMISSION LOSS CEILINGS

High transmission loss ceilings can attenuate sound which is emanating from upper floors. Such a ceiling is illustrated in Figure 7.14. Two-pound sheet lead is normally employed that can be fastened to board using staples or adhesives before installation. The sheets should be lapped at least 2 inches and carried up the perimeter walls at least 4 inches. The perimeter should be thoroughly caulked to obtain an air-tight seal. Also, at each point of suspension isolation hangers should be placed. Last, the ceiling slab above should be covered with a fibrous absorptive material.

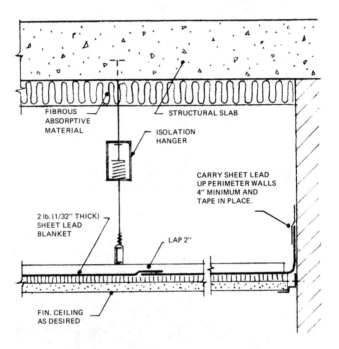

Figure 7.14 High transmission loss ceiling.

NOISE BARRIERS

Figures 7.15 through 7.31 provide illustration of assemblies, partitions, and laminates in sequence according to thickness. Each is given a numerical rating of Sound Transmission Class (STC). Table 7.2 (p. 112) gives an idea of the effectiveness of each. A transmission loss curve at various frequencies is also given for each.

ENCLOSURES, SHIELDS AND BARRIERS—DESIGNING WITH LEAD

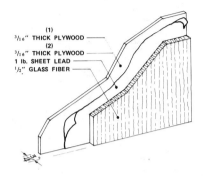

(1) 3/16" THICK PLYWOOD
(2) 3/16" THICK PLYWOOD
1 lb. SHEET LEAD
1/2" GLASS FIBER

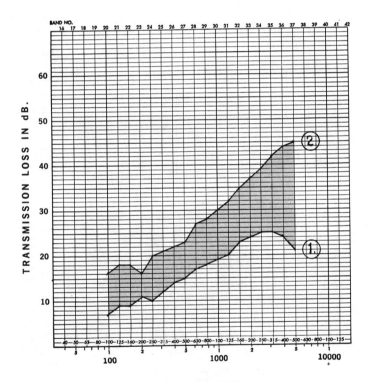

CURVE NO.	CONSTRUCTION	STC
Curve #1	3/16 inch thick plywood alone	19
Curve #2	3/16 inch thick plywood, 1 lb. sheet lead, 1/2 inch glass fiber, adhesive laminated.	29

Figure 7.15

96 INDUSTRIAL NOISE CONTROL HANDBOOK

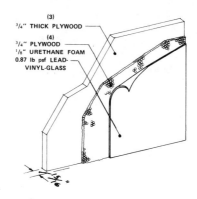

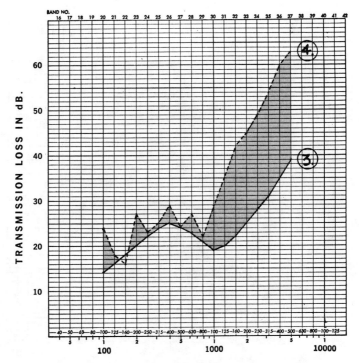

CURVE NO.	CONSTRUCTION	STC
Curve #3	¾ inch thick plywood alone	22-24
Curve #4	¾ inch thick plywood, ⅛ inch thick urethane foam, .87 psf. lead/vinyl/glass material.	29

Figure 7.16

ENCLOSURES, SHIELDS AND BARRIERS—DESIGNING WITH LEAD

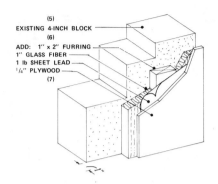

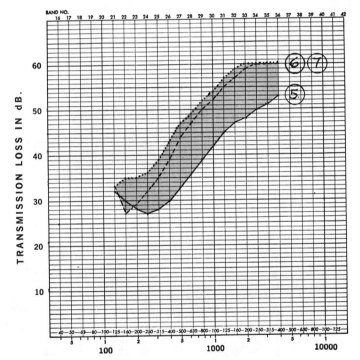

CURVE NO.	CONSTRUCTION	STC
Curve #5	Existing 4 inch thick cinder block wall	38
Curve #6	Add: 1" x 2" furring, lightweight glass fiber in cavity, ¼ inch thick plywood.	45
Curve #7	Same as #6 except 2" x 2" furring instead of 1" x 2" furring.	49

Figure 7.17

98 INDUSTRIAL NOISE CONTROL HANDBOOK

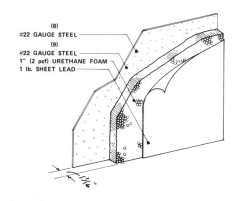

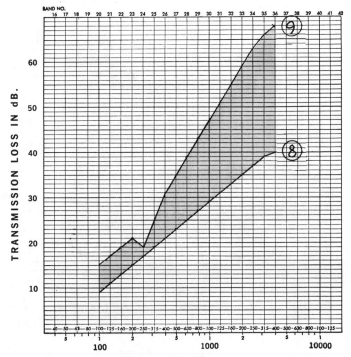

CURVE NO.	CONSTRUCTION	STC
Curve #8	#22 Gauge Steel Sheet alone	27
Curve #9	#22 Gauge Steel Sheet, 1 inch thick (2 pcf.) urethane foam, 1 lb. sheet lead.	34

Figure 7.18

ENCLOSURES, SHIELDS AND BARRIERS—DESIGNING WITH LEAD

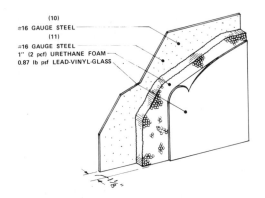

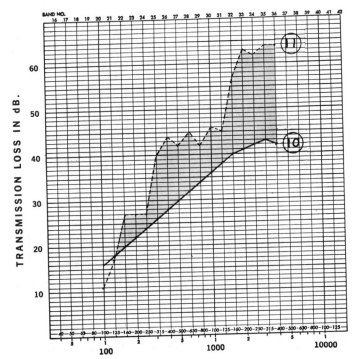

CURVE NO.	CONSTRUCTION	STC
Curve #10	#16 Gauge Steel Sheet alone	34
Curve #11	#16 Gauge Steel Sheet, 1 inch thick (2 pcf.) Urethane foam, .87 psf. Lead/Vinyl/Glass material.	41

Figure 7.19

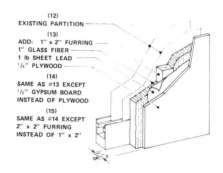

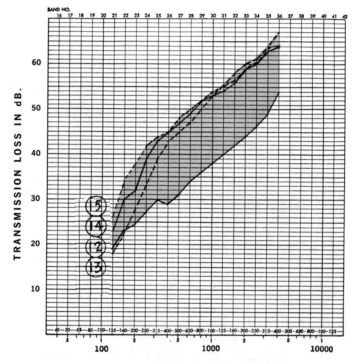

CURVE NO.	CONSTRUCTION	STC
Curve #12	Existing Standard Steel Office Partition	36
Curve #13	Add 1" x 2" furring, 1 inch thick glass fiber, 1 lb. sheet lead, ¼ inch thick plywood.	42
Curve #14	Same as #13 except ½ inch thick gypsum board instead of ¼ inch thick plywood.	47
Curve #15	Same as #14 except 2" x 2" furring instead of 1" x 2" furring.	50

Figure 7.20

ENCLOSURES, SHIELDS AND BARRIERS—DESIGNING WITH LEAD

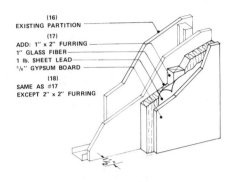

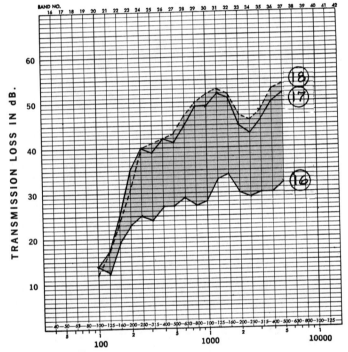

CURVE NO.	CONSTRUCTION	STC
Curve #16	Standard steel stud and gypsum board construction (not calked).	29
Curve #17	Add 1" x 2" furring 24" on center, bedded on calking beads, 1 inch thick glass fiber, 1 lb. sheet lead, 5/8 inch thick gypsum board.	39
Curve #18	Same as #17 except 2" x 2" furring instead of of 1" x 2" furring.	41

Figure 7.21

102 INDUSTRIAL NOISE CONTROL HANDBOOK

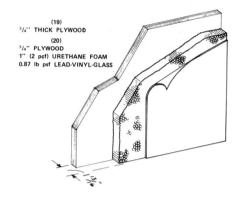

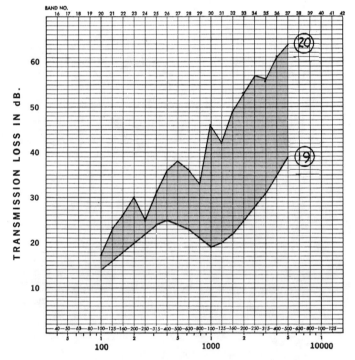

CURVE NO.	CONSTRUCTION	STC
Curve #19	¾ inch thick plywood alone	22-24 (Est.)
Curve #20	¾ inch thick plywood, 1 inch thick (2 pcf.) urethane foam, .87 psf. lead/vinyl/glass material.	40

Figure 7.22

ENCLOSURES, SHIELDS AND BARRIERS—DESIGNING WITH LEAD 103

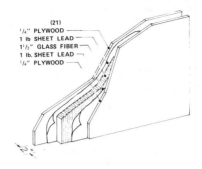

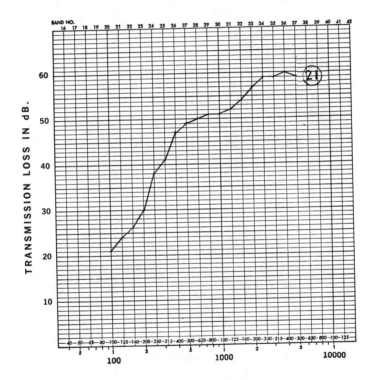

CURVE NO.	CONSTRUCTION	STC
Curve #21	¼ inch thick plywood, 1 lb. sheet lead, 1½ inch (2 pcf.) glass fiber, 1 lb. sheet lead, ¼ inch thick plywood, 1⅝ inch wood perimeter frame.	47

Figure 7.23

104 INDUSTRIAL NOISE CONTROL HANDBOOK

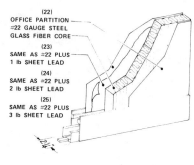

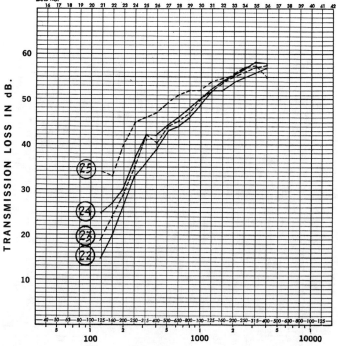

FREQUENCY IN HERTZ

CURVE NO.	CONSTRUCTION	STC
Curve #22	Standard Steel Office Partition, #22 gauge steel, glass fiber core.	38
Curve #23	Same as #22 except add 1 lb. sheet lead to one finish face.	43
Curve #24	Same as #22 except add 2 lb. sheet lead to one finish face.	46
Curve #25	Same as #22 except add 3 lb. sheet lead to one finish face.	52

Figure 7.24

ENCLOSURES, SHIELDS AND BARRIERS—DESIGNING WITH LEAD

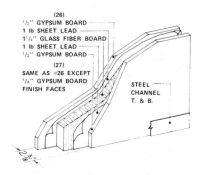

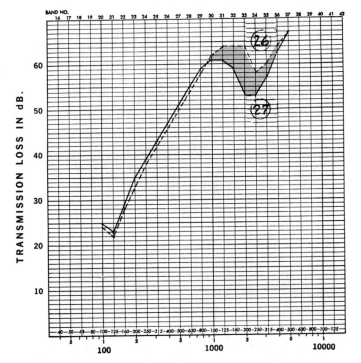

CURVE NO.	CONSTRUCTION	STC
Curve #26	½ inch thick gypsum board, 1 lb. sheet lead, 1¼ inch thick (6 pcf.) glass fiber board, ½ inch thick gypsum board, 1 lb. sheet lead, steel channel runners top and bottom.	46
Curve #27	Same as #26 except ⅝ inch thick gypsum board instead of ½ inch thick gypsum board.	47

Figure 7.25

106 INDUSTRIAL NOISE CONTROL HANDBOOK

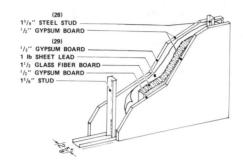

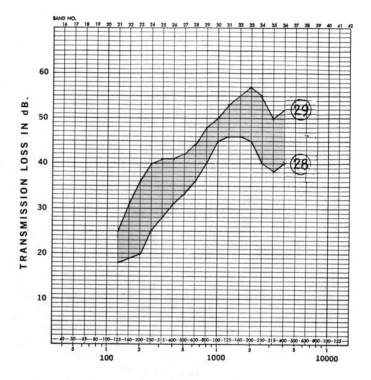

CURVE NO.	CONSTRUCTION	STC
Curve #28	½ inch thick gypsum board, 1 ⅝ inch steel studs, ½ inch thick gypsum board.	36
Curve #29	½ inch thick gypsum board, 1 lb. sheet lead, 1 ½ inch thick glass fiber board, ½ inch thick gypsum board, 1 ⅝ inch steel studs.	47

Figure 7.26

ENCLOSURES, SHIELDS AND BARRIERS—DESIGNING WITH LEAD 107

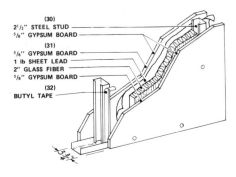

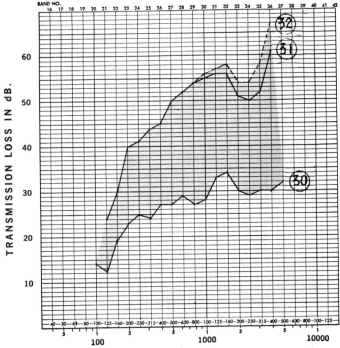

CURVE NO.	CONSTRUCTION	STC
Curve #30	⅝ inch thick gypsum board, 2½ inch steel studs, ⅝ inch thick gypsum board, (not calked).	29
Curve #31	⅝ inch thick gypsum board, 1 lb. sheet lead, 2 inch thick (2 pcf.) glass fiber blanket, ⅝ inch thick gypsum board.	48
Curve #32	Same as #31 except butyl tape (½" x 1/16") used between studs and finish materials.	48 (Est.)

Figure 7.27

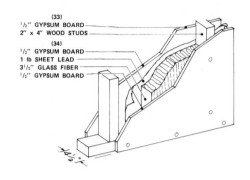

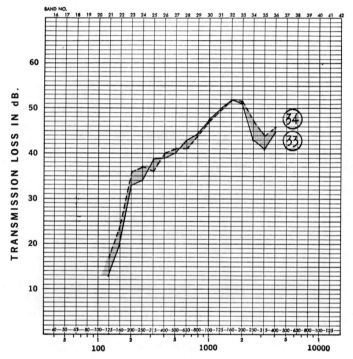

CURVE NO.	CONSTRUCTION	STC
Curve #33	½ inch thick gypsum board, 2″ x 4″ wood studs	37
Curve #34	½ inch thick gypsum board, 1 lb. sheet lead, 3½ inch thick glass fiber, ½ inch thick gypsum board.	41

Figure 7.28

ENCLOSURES, SHIELDS AND BARRIERS—DESIGNING WITH LEAD 109

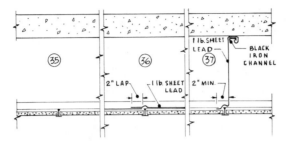

SUSPENDED ACOUSTICAL CEILING

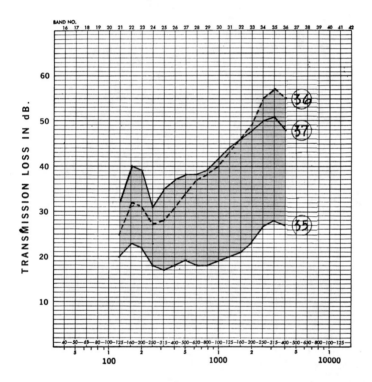

FREQUENCY IN HERTZ

CURVE NO.	CONSTRUCTION	STC
Curve #35	Standard suspended ceiling construction, "T" bar grid, low density ceiling panels.	21
Curve #36	Same as #35 except add 1 lb. sheet lead blanket lapped 2 inches.	36
Curve #37	Same as #35 except add 1 lb. sheet lead vertical barrier hung from slab over partitions.	41

Figure 7.29

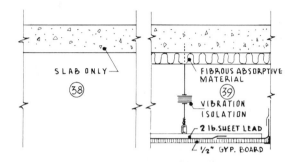

MECHANICAL ROOM CEILING

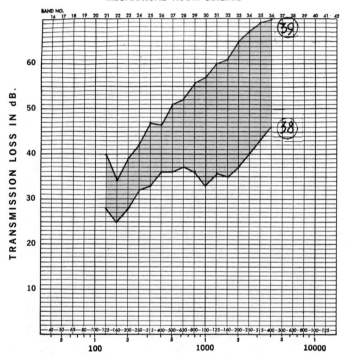

FREQUENCY IN HERTZ

CURVE NO.	CONSTRUCTION	STC
Curve #38	Typical concrete slab.	34
Curve #39	½ inch thick gypsum board ceiling hung on vibration isolators, absorptive material adhered to slab, 2 lb. sheet lead blanket, lapped 2 inches, over ceiling construction.	52

Figure 7.30

ENCLOSURES, SHIELDS AND BARRIERS—DESIGNING WITH LEAD

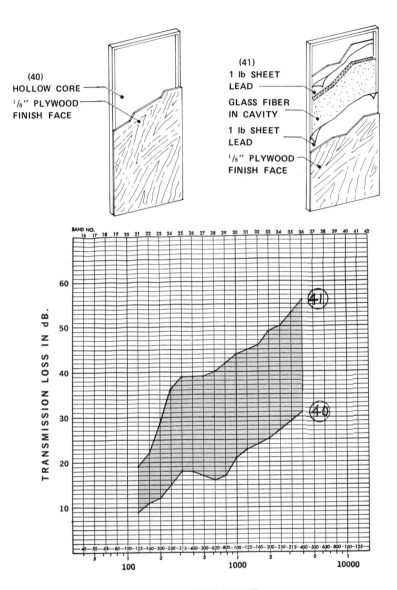

CURVE NO.	CONSTRUCTION	STC
Curve #40	Standard hollow core plywood door.	21
Curve #41	Same as #40 except 1 lb. sheet lead laminated to both inner finish faces, glass fiber in cavity.	42

Figure 7.31

Table 7.2 STC Key

STC Number	Effectiveness
25	Normal speech that can be easily understood
30	Loud speech that can be well understood
35	Loud speech that is audible but not intelligible
42	Loud speech that is as audible as a murmur
45	Ear must be strained to hear loud speech
48	Loud speech that is barely audible
50	Loud speech that cannot be heard at all

SEAMS

In fastening lead sheets together when used on piping, ductwork and other such applications, three basic types of seams are air-tight and can regulate noise. These seams are standing, flat and batten. All three types are illustrated in Figure 7.32. Flat seams are employed to fasten wall panels to each other and also to fasten roof panels together. This type of seam can easily be hidden. Standing seams are commonly used for roofing and flashing. This type of seam is similar to the flat seam, except that it protrudes about $1^1/_2$ inches. Batten seams are used for corner and roof-to-wall seams.

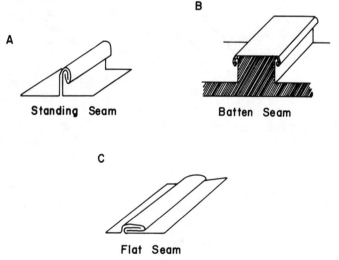

Figure 7.32   Three types of seams used in barrier installation.

LONG ISLAND LIGHTING COMPANY'S
NORTHPORT, NEW YORK PLANT—CASE HISTORY

This electrical-generating plant site is surrounded by a quiet residential community. The plant had a problem—it wanted to increase its capacity without increasing the level of noise normally generated. This problem was met head-on by enclosing the new equipment with lead panels. After these panels were installed and the new equipment put into operation, the plant was very quiet and did not disturb the surrounding residential community.

Original design work of the three units of the plant showed much concern for creating a low dB level in the surroundings. These units required equipment which would create much greater noise levels than those which would be acceptable. And, as an added problem, the generated noise would travel further in the quiet residential surroundings. Therefore, much optimization was made in selection of equipment. Original equipment had the lowest practical sound level built in. All of the equipment with high noise levels was enclosed by sound-absorbing materials, and soundproof walls were installed in the entire plant. However, problems arose when the new equipment was installed in the first unit. A fan with its essential ductwork was too large to fit in the particular plant area, so the existing acoustic walls would not attenuate the sound of the new equipment. Noise problems were determined to be in the ductwork. After careful consideration, lead was found to be the most efficient and practical material to be used in solving the problem. Panels of lead with four other materials were placed at strategic locations throughout the system. These panels consisted of lead sandwiched between mineral wood for insulation, and on the exterior side was aluminum or steel. On the interior side of the panel (the side exposed to the noise) wire mesh was installed. Either one or two layers of lead was used in the panel, depending on the level of noise in the area in which the panel was to be placed.

As soon as the acoustical systems were installed in units one and two of the plant, the third unit posed very little problem; its noise was deadened by a system of panels and barriers similar to those used in the first two units. All of the panels in the plant were used to enclose the ductwork, fan and precipitator of unit three. Precut panels of three types were obtained for installation. Since the panels were precut, installation time and cost were conserved, allowing rapid setup of the third unit for testing. About 73,000 square feet of lead was employed in this sound system. Most of this lead was used in unit three.

Figure 7.33 shows a cut-away of one of the panels used in the plant. Figure 7.34 illustrates the noise control system employed in unit three of the LILCO plant.

Figure 7.33 Frank J. Wells, LILCO Electrical Engineering Staff, displays cut-away of one of the two types of thermal/acoustical panels that solved major noise problems at the company's Northport plant. The panel contains a 1-pound sheet lead sound barrier, seen extending from between two layers of mineral wool, which provides the panel with thermal insulating properties. Some of the equipment covered by the panels can be seen in the background.

ENCLOSURES, SHIELDS AND BARRIERS—DESIGNING WITH LEAD 115

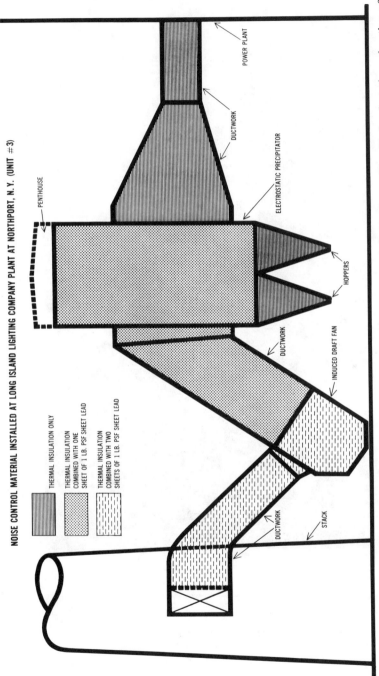

Figure 7.34 Line drawing of power generation unit 3 at Long Island Lighting Company's Northport, New York, plant shows location of three types of composite material building panels used to solve noise control problems for the installation's exterior ductwork and accessory equipment.

Figure 7.35 View of combination thermal/acoustical panel being installed to silence equipment at unit 3. Ductwork leading from induced draft fan equipment to unit's 600-foot-high stack is illustrated.

Figure 7.36 Continuation of panels covering additional ductwork leading to stack, panels covering induced-draft fan housing (bottom, center), and ductwork from precipitator to fan (right center). Lower walls of precipitator housing (partly visible at bottom right) also were enclosed with panels. Emission stack and other equipment for unit 2 are visible in background.

REFINERY NOISE REDUCTION—CASE HISTORY

Oil refineries and chemical plants encounter problems in complying with OSHA standards on noise control. Installing noise reduction systems for large plant equipment is expensive and causes downtime for the equipment, thus decreasing production. One east coast refinery, however, has proved that sound attenuation systems can be installed without changing the normal plant schedule. This plant reduced its noise levels to meet OSHA standards without disturbing plant operations by installing a combination of glass fiber sheet lead and aluminum on its equipment.

The consulting firm that designed the acoustic system—Donley, Miller and Nowikas, Upper Montclair, New Jersey—pinpointed the major plant equipment noise problems in blowers, gearboxes, steamchests and several sections of the large pipe system inside the plant. Some of the essential measures needed were shrouding the noisy equipment in thick layers of glass fiber in conjunction with thinner layers of sheet lead, and then jacketing the noise shroud in heavy-gauge aluminum.

To silence the gearboxes, anchor pegs were attached to the gearbox casings, which supported 1.5-2.0-inch thick layers of fiberglass material. These anchor pegs were bonded rather than welded because welding would have created a spark hazard in an area containing many flammable substances. Three and 6 pound per cubic foot (pcf) fiberglass was installed by Parkway Insulation Company, Hackensack, New Jersey. The fiberglass was then covered with sheet lead, most of which was 2 lb per ft^2 (psf) at 0.0312-inch thick; however, some psf at 0.0516-inch thick sheet lead was also used.

Noise reduction on some parts of machinery such as valve handles and gauges was achieved by adding silicone rubber and aluminum sliding doors. Silicone rubber used on the valve handles with the sliding aluminum doors covering them eliminated the noise produced there. Aluminum doors in conjunction with clear plastic stopped noise around various gauges in the plant. Blowers, another potential noise source, were enclosed to control their noise. The blowers were first layered with glass fiber and sheet lead and were then covered with sound-insulated outer panels of aluminum, glass fiber, and lead. These panels were movable and provided sufficient access to the equipment when required. The noisy piping in the refinery was wrapped in glass fiber, sheet lead and aluminum jacketing.

The refinery revealed that the cost of the sound attenuation system for the large blower housing was under $10,000, and the cost for the gearboxes was under $6,000.

ENCLOSURES, SHIELDS AND BARRIERS—DESIGNING WITH LEAD 119

Figure 7.37 Sheet lead, glass fiber and aluminum reduced sound levels near this blower by more than eight decibels.

Figure 7.38 Gearbox noise was reduced to levels permitted by OSHA.

Figure 7.39 Noise problem areas such as this large steamchest were pinpointed by the acoustical consulting firm.

SPINNING MILL NOISE REDUCTION[2]

High noise levels occur in textile plants and a large source of this noise is the pin-drafting area. An industrial noise control program instituted in a southern textile plant proved to be effective in noise attenuation in this area.

The spinning mill is typical of most plants where employees in the drafting areas are exposed to high levels of noise throughout the day. The program was to reduce the noise to levels acceptable to the Occupational

ENCLOSURES, SHIELDS AND BARRIERS—DESIGNING WITH LEAD

Safety and Health Workplace Noise Standard in an economically feasible way. The program was also designed so that similar plants could institute it if successful. The program had two phases. Phase one was an application of acoustical absorption to the mill room. Phase two employed machine treatment of the pin-drafting equipment. The machinery in the carding and drafting area, typical of most yarn-spinning mills, is illustrated in Figure 7.40. From the blenders and cards where the crude fiber is combed and blended, the product flow proceeds to the servo-drafters and pin-drafters where the sliver is drawn and combed. In a different section of the plant this sliver is then spun into yarn and wound on bobbins by the spinning frames.

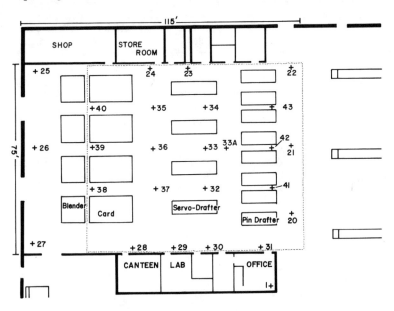

Figure 7.40 Layout of machinery in the noise problem area.

Noise levels near the pin-drafters often exceeded 100 dBA. Equipment investigation revealed that the blenders, servos and cards were acceptably quiet when in operation. The high noise levels in this area were due rather to the high reverberant field noise created by the pin-drafters and spinning frames. It was decided that acoustical absorption be applied in the pin-drafting and servo areas, since reverberant field sound can be restrained by acoustical treatment of the room. A noise attenuating system was tested on one of the pin-drafters to determine the amount of machine noise reduction and to develop a prototype that could be applied to the other seven pin-drafters.

Table 7.3 presents the data that resulted from the test, both before and after noise situations. Sampling by tape recording and third-octave analysis revealed noise spectra data at positions 27 and 33, which is illustrated in Figure 7.41. In position 33A a B&K Model 166 Environmental Noise Classifier was installed to determine the time domain relationship of various noise levels. A number of eight-hour samplings resulted in an average exposure level of 220%, based on the 90 dBA for an eight-hour exposure of 100%.

Table 7.3 Sound Levels in the Carding and Drafting Area[3]

Process	Position	Sound Levels Before (dBA)	Sound Levels After (dBA)	Reduction (dBA)
Pin-drafting	20	93.0	92.0	1.0
	21	93.5	93.0	0.5
	22	94.0	91.5	2.5
	31	92.5	91.0	1.5
	41	96.5	96.0	0.5
	42	97.0	95.5	1.5
	43	98.5	98.3	0
	Ave	95.0	93.9	1.1
Servo-drafting	29	90.0	86.0	4.0
	30	92.5	88.0	4.5
	32	93.0	90.0	3.0
	33	93.5	90.0	3.5
	34	93.0	90.0	3.0
	23	90.5	86.5	4.0
	24	90.5	85.5	5.0
	35	90.0	89.5	0.5
	36	90.0	90.0	0
	37	91.0	89.0	2.0
	Ave	91.4	88.5	3.0
Blending/carding	28	90.0	85.0	5.0
	38	88.5	84.0	4.5
	39	89.0	85.5	3.5
	40	88.0	85.0	3.0
	25	85.5	81.0	4.5
	26	86.0	83.5	2.5
	27	84.0	80.0	4.0
	Ave	87.3	83.5	3.9
Office area	1	65.0	59.0	6.0

ENCLOSURES, SHIELDS AND BARRIERS—DESIGNING WITH LEAD

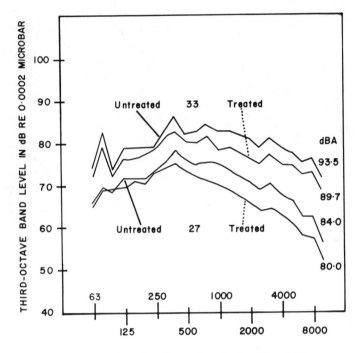

Figure 7.41   Test data taken at sites 27 and 33.

In the phase one treatment consideration was given to the application of unit absorbers like baffles, panels and cylinders in the drafting and carding area. However, these unit absorbers were rejected by the plant personnel because they created an area where lint would accumulate. Plant personnel also rejected a continuous acoustical ceiling because of the high cost of modifying lights, ductwork and sprinklers, modifications essential for the installation.

To meet this plant's noise problems, functional absorbers were designed that would not interfere with ducts, sprinklers and lights when installed. These special absorbers are illustrated in Figure 7.42. The absorbers use vinyl-covered acoustical panels, shaped so that they can be easily vacuumed. They also will not build up any surface static charge and thus will not attract any lint. Table 7.4 shows the acoustic efficiency of the absorbers.

The absorbers were installed in rows five feet on center at nine feet above floor level. They covered an area 75 x 100 feet above the servo-drafting and pin-drafting area. This total area is enclosed by a dashed line in Figure 7.40. About 700 square feet of the north wall near the

Figure 7.42 These specially designed functional noise absorbers minimize housekeeping problems and do not require relocation of air ducts or sprinklers. The absorbers utilize special vinyl-covered acoustical panels shaped in such a way that they can be easily vacuumed.

servo-drafters and pin-drafters was coated with a vinyl-covered acoustical wall system that reduced sound reflection, thus further attenuating sound.

As a result of this system the functional absorbers and wall panels added approximately 4300 sabins, measured at 500 Hertz, to the test area. The untreated area had contained 1720 sabins of absorption, which was due mainly to the fiber being processed there. Therefore, there was a 3.50-fold increase in absorption after treatment of 6020 sabins of the area.

Table 7.4 Acoustical Performance of Function Absorbers
Sabins per Lineal Foot

Hertz							
125	250	500	1000	2000	4000		4 Freq. Av.
0.22	1.22	3.09	5.48	4.56	3.29		3.6

ENCLOSURES, SHIELDS AND BARRIERS—DESIGNING WITH LEAD

Using this data it was estimated that the reverberant field sound pressure level would decrease by 5 dB. Table 7.3 indicates that the 5 dB reduction in the field was indeed obtained. Noise levels in the servo-drafting area were reduced to 90 dBA and lower. After treatment, dosimeter readings at position 33A averaged 119% exposure for eight hours rather than the 220% before treatment. In addition, the noise levels in the offices of the plant were reduced by 6 to 8 dB.

The pin-drafting machines were in fact the major source of noise in the textile installation because of the metal-to-metal contact between bars in the faller bar area. A machine operator would receive very little protection from the loud sound levels by the machine casing. Figure 7.43 illustrates the noise levels in dBA at various locations around the pin-drafter. The noisiest location is #1 which is directly over the faller bars.

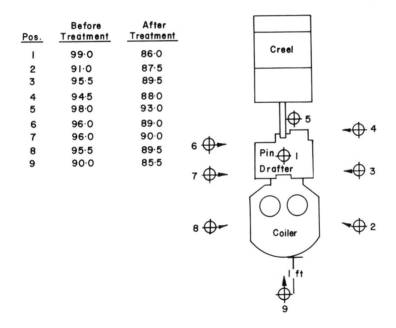

Pos.	Before Treatment	After Treatment
1	99.0	86.0
2	91.0	87.5
3	95.5	89.5
4	94.5	88.0
5	98.0	93.0
6	96.0	89.0
7	96.0	90.0
8	95.5	89.5
9	90.0	85.5

Figure 7.43 Acoustical data taken around pin-drafter area.

As a method of reducing the loud noise the faller bars were coated with plastic. However, reduction averaged about 1.5 dBA. Next the sound path from the machine was blocked. A hinged cover was designed to fit over the top of the faller bar area. Since the sides of the machine were open, metal panels were installed. These side panels and the cover were damped with viscoelastic vibration damping material with acoustical foam one inch thick. This total acoustical system reduced the noise levels around the machine by 6.4 dB, down to a level of 88.7 dBA. Actual test data can be seen in Figure 7.43. The third-octave plot of the sound spectrum before and after the treatment is illustrated in Figure 7.44. When all eight pin-drafting machines were treated, noise levels in that area of the plant were reduced to 90-92 dBA.

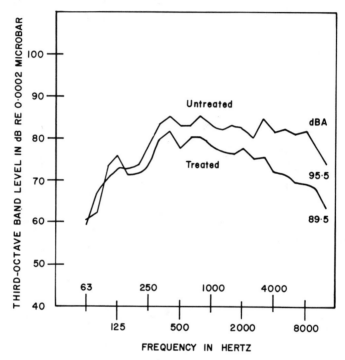

Figure 7.44 Third-octave plot of the sound spectrum before and after treatment.

NOISE ATTENUATION IN TURBINE-POWERED NATURAL GAS COMPRESSOR STATIONS

When natural gas compressor stations have to be situated near residential areas, the high levels of noise from the turbine compressors can prove to be troublesome. Many design problems arose when two Canadian natural gas companies encountered this situation. The two utilities—Ontario Division of Northern and Central Gas Corp., Ltd. and the Union Gas Company of Canada Ltd.—were determined to develop a noise attenuation system to reduce the 115 dB turbine noise to acceptable limits. Northern and Central had made plans for construction of a compressor station powered by two 1000-hp turbines at Hagar, in Northern Ontario. Union Gas had planned a larger installation at Lobo, Ontario, which would be powered by one 12,500-hp turbine.

Instead of installing noise attenuating systems in the plants after they had been put into operation, the two companies worked together to solve potential noise problems while the plants were still in the design stages. The result of their efforts was a composite acoustical building that would decrease the turbine noise to acceptable levels. The acting consultants in the designing of the stations were International Pipeline Engineering Ltd. (IPEL), a subsidiary of Trans Canada Pipelines, Ltd. For the compressor stations IPEL had drawn up strict specifications that called for noise controls to be built into the turbine equipment and the structure surrounding them.

The construction of the compressor stations was done by Pro-Eng Buildings, Ltd., Ontario, who developed an acoustical building panel to house the stations. The final design of the panel had an inside face of 2 pounds per square foot (0.312-inch thick) sheet lead bonded to the back side of 24-gauge (0.0239-inch thick) V-rib galvanized steel sheet. The center core of the panel consisted of one 4-inch thick and one 3-inch thick piece of low density fiberglass layer with a vinyl backing. The exterior face of the panel was 26-gauge (0.0179-inch thick) galvanized steel sheet with a baked enamel finish. The panel also used girts that were cold-rolled from steel Z-sections. A section of this panel is illustrated in Figure 7.45.

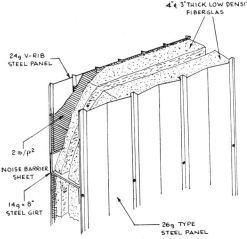

Figure 7.45 Cut-away view of new acoustical building panel shows position of 2 pounds per square foot ($1/32$-inch thick) sheet lead that is bonded to back face of panel.

When tested for noise control quality, this special building panel obtained a Sound Transmission Class (STC) rating in excess of 40 dB. The panel was applied both in roofing and siding in the two companies. The 12,500-hp unit at Union Gas Corp. contained in a 43 x 57 x 22 foot building made of these panels. The two 1000-hp units at Northern and Central are housed in a 30 x 50 x 14 foot building. According to IPEL and pipeline operators, there have been no complaints of noise in the residential areas near the companies.

ENCLOSURES REDUCE CONSTRUCTION SITE NOISE

A step forward in the reduction of construction site noise in a heavily populated area was achieved by noise control engineers for the Department of Environmental Control, City of Chicago. In collaboration with a manufacturer of noise control systems, the engineers demonstrated and tested a prototype sound-absorbent enclosure at a water and sewer department work site on a main thoroughfare. The results of sound level meter tests showed that the enclosure not only reduced the construction noise level but was also particularly effective at those high frequencies known to be most damaging to the human ear and also most annoying to the city inhabitants.

In testing the enclosure, several city department workmen cut a 12-inch diameter, ductile iron water main with a gasoline powered cutting saw while engineers measured sound pressure levels at varying distances from the cutting site. At ten feet from the site, the cutting operation recorded a 97 dBA reading without the enclosure and a 12 dB reduction to 85 dBA with the closed curtain in place. The ambient noise level in the area was 76 dBA. A spokesman for the department said that the enclosure was an effective means of reducing construction noise. Engineers commented that the 12 dB reduction could be expected in most work situations involving comparable size construction power tools. They also noted that the reduction was equivalent to moving the noise source four times as far away from a common point.

The four-sided enclosure with separate top from Singer Partitions, Inc., Chicago, was constructed using six free-standing columns to support a framework of overhead roller curtain track (see Figures 7.46 to 7.48). Lead-filled, vinyl-coated fiberglass-grommeted curtain material with an inner facing of sound-absorbent acoustic foam was hung from rollers on the track. The same material was used for the roof section to effect a tight closure. Two men installed the enclosure in 45 minutes. The estimated cost of the 9-foot long by 7-foot high by 6-foot wide enclosure with hardware was $1500. It could be made to be movable for such

ENCLOSURES, SHIELDS AND BARRIERS—DESIGNING WITH LEAD 129

Figure 7.46 Construction noise abatement, city of Chicago. Framing for sound-absorbent enclosure consists of 6 free-standing columns supporting heavy duty roller curtain track. Ductile iron water main (12-in. diameter; $1/2$-in. wall), foreground, was cut at bell end for sound readings.

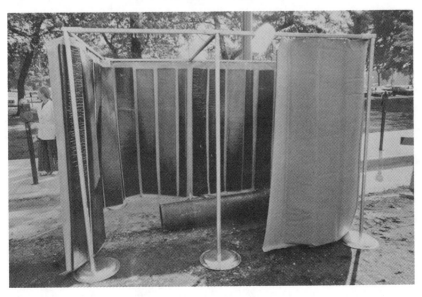

Figure 7.47 Construction noise abatement, city of Chicago. Partially erected enclosure of leaded-vinyl fiberglass material hung on trolley rollers from track. For sound absorbency, work side of enclosure is faced with 10-in. wide panels of acoustic foam with Mylar-covered surface, which prevents foam pores from clogging.

Figure 7.48 Construction noise abatement, city of Chicago. Full enclosure has lapped roof of leaded-vinyl fiberglass material to effect tight seal. Vertical edges of curtain have Velcro strips for easy entry and exit.

operations as pavement breaking and with openings for ventilation equipment and power lines.

ENCLOSURE REDUCES HYDRAULIC PUMP NOISE

The absorption and containment of high-frequency noise that disturbed and distracted precision machine equipment operators was reported by West Milton Precision Machine & Tool, Inc. To identify the noise source, company officials commissioned a noise survey and octave band analysis by a Dayton noise control consultant, H. Routson Associates. The survey indicated that the majority of the noise in the precision production area was being generated by an off-floor, 10-hp hydraulic pump, cooling fan, related valving and pressures within the fluid lines.

The octave band analysis showed that the noise levels were predominantly in the higher-frequency ranges. A total enclosure that could be installed easily and that offered ready accessibility for pump maintenance was recommended as a feasible solution to the noise problem.

A four-sided enclosure with separate top from Singer Partitions, Inc., Chicago, was constructed with tubular steel frames supported by platform legs (Figure 7.49). The framing was covered with leaded-vinyl curtain material with an inner facing of sound-absorbent foam panels. Two men installed the complete enclosure in 45 minutes. The leaded vinyl-foam combination effectively reduced the noise level inside the enclosure with a consequent reduction of noise in the adjacent precision machining area. Sound level readings before and after in the machining area were 96 and 78. The increase in employee morale and efficiency is such that the company has planned additional installations of this type.

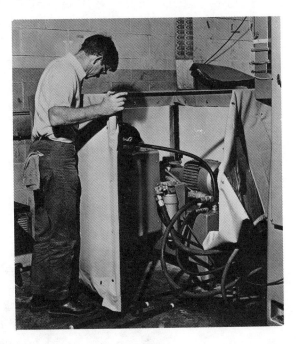

Figure 7.49 Easily installed total enclosure with framing covered by leaded-vinyl curtain material.

NOISE-ATTENUATING CURTAINS

Can music lessons and basketball practice be successfully conducted side-by-side in a school gymnasium? This question was posed when the junior high school in a Michigan community converted its gymnasium stage into a much-needed music room, closed off from the gym floor by folding wood doors. The effectiveness of the plan for dual-purpose use of the space, though logical and obviously economical, was seriously threatened when exuberant shouts of active physical education classes continually interfered with band practice and made concentration difficult for the school's choral group.

Looking for a means to ensure a quiet environment for music students, administrators of the Nellie B. Chisholm junior high school in Montague, Michigan, turned for advice to a firm of acoustical engineers in nearby Grand Rapids. To retain use of the new rehearsal room for 100 students in the band program and the 35-member chorus, a survey by sound engineers of Leggett-Michaels Co. recommended floor to ceiling installation of movable Sound Stopper curtain, a type that is specially fabricated to absorb and suppress high-frequency noise (Figures 7.50 to 7.52). The approved installation involved hanging ten curtains, each 19 feet high by 4 feet wide, from a steel track with ends curved to accommodate complete stage-wide opening of the curtains. The track itself was positioned in a bulkhead built into the ceiling. Secure closure was achieved with fabric fastener tapes, sewn into the edge of each curtain. The curtain is made of a leaded-vinyl fabric to which 10 inch-wide vertical strips of flexible acoustical urethane foam are permanently affixed. The strip arrangement permits the sound-smothering curtain to fold in accordion-pleted style for opening when the 60- by 24-foot area is needed as a stage for school functions.

Figure 7.50 Sound enclosure curtain (partially open at left) is closed to seal off noise from physical education classes.

ENCLOSURES, SHIELDS AND BARRIERS—DESIGNING WITH LEAD 133

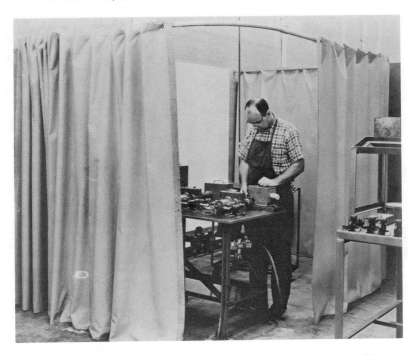

Figure 7.51 Industrial firms and institutions with on-premise noise problems are making increasing use of sound-suppressing curtains like the enclosure shown, located at Signode Corporation, Glenview, Illinois, where it effects an estimated 50% reduction in noise from air motors being tested before assembling into the company's steel-strapping machines. The high-pitched whine of the air motor, although not especially high on the decibel scale, is distraction for production workers in the adjacent production area. The approximately 8- by 10-foot space shown here is enclosed by the single-thickness curtain, an acoustically engineered woven fiberglass cloth coated both sides with lead-filled vinyl. In higher-intensity sound-suppressing problems, a ceiling of the same material is mounted on a double-truss roof-structure, or, where practicable, the curtain is hung from the building's ceiling, in both cases providing a total enclosure. Double curtains can be used where additional reduction is required. The curtain is a tough, durable material weighing $4^{1}/_{2}$ pounds per square yard, is beige colored, easily washable, and moves freely on 4-wheel ball bearing rollers. The curtain closes tightly through use of a self-adhering nylon fastening strip seen on the left-hand curtain.

QUIETING NOISY WASHER-DRYER SYSTEMS

Industrial washer-dryer systems are often excessively noisy, with most noise occurring in the dryer section from rotating blower blades and high-pressure blow-off systems. ACON, Inc., has offered two approaches for attenuation of dryer noise:

134 INDUSTRIAL NOISE CONTROL HANDBOOK

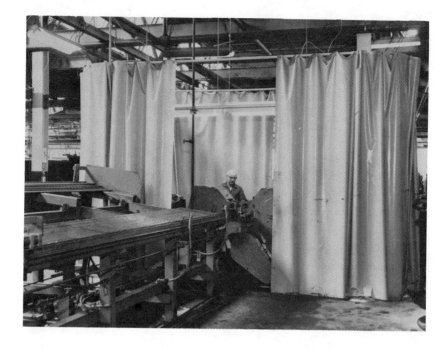

Figure 7.52 The loud noise from this cutoff machine at Standard Screw Company, Elyria, Ohio, was effectively suppressed by adoption of the curtain shown here.

- total enclosure of the washer and blow-off plenum
- separate treatment of each noise source and transmission path

The latter is the most practical. Typically, noise levels produced by dryer installations are between 95 and 107 dBA. With treatment, these levels can be reduced to between 80 and 87 dBA. The system is shown in Figure 7.53.

OFFICE SILENCE WITH LEAD

To avoid the crush of subway crowds, overloaded buses, late commuter trains, take an elevator to work. This mode of traveling is a reality at the John Hancock Center Building in Chicago, Illinois. The 1,127-foot combination office-residential building, the fourth tallest building in the world, contains a sky lobby and 705 apartments in its uppermost 48 floors. Located on North Michigan Avenue, the Hancock Center Building was designed by the architectural firm of Skidmore, Owings & Merrill of

ENCLOSURES, SHIELDS AND BARRIERS—DESIGNING WITH LEAD 135

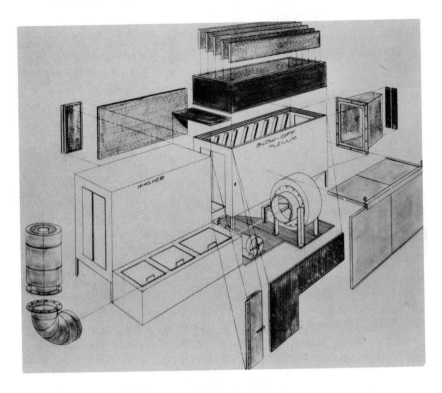

Figure 7.53 Noisy washer-dryer system.

Chicago. The building is one of the first to combine offices, retail stores, restaurants, residences and parking garages in one structure.

An important factor in planning a multiple-use building of this type is the attenuation of noise. Noise generated in machinery rooms located throughout the building for water pump boosters, elevator motors and other necessary building service equipment is partially screened during the day by the high level of office machine and street noise. In the evening, however, when street noise has diminished and the only noise coming from the offices and restaurant areas is that of clean-up crews, sound emanating from these machinery rooms located above and below the living areas is more noticeable. Added to this is the critical need to maintain quiet between apartments on the residential floors. A potential noise problem was solved through the joint efforts of the architect, the acoustical consultants, Bolt, Beranek & Newman, Inc., of Cambridge, Massachusetts, and the dry wall contractor, McNulty Brothers Co., of Chicago.

Two factors had to be considered in the design of the sound barriers for the apartments: the movement of the building in the wind and the narrow width of the mullions. In the construction of the apartment walls, two pieces of $5/8$-inch gypsum board were used on either side of $3 5/8$-inch metal studs. The void between the gypsum board was filled with acoustical wool. This construction provided a sound transmission class (STC) of 45. The walls were carried to within 28 inches of the perimeter walls. At that point, the $4 7/8$-inch thick wall was narrowed to $1 7/8$ inches to match the width of the mullion.

A unique design made it possible to bridge the gap between the interior wall and the mullion. A channel 2 inches deep and $7/8$ inch wide was attached to the mullion as well as to the floor and the spandrel overhead. A channel was also attached to the end of the interior wall to provide for a slip joint to compensate for the movement of the building. Gypsum board, $5/8$-inch thick, was fitted inside the channel allowing a $1/4$-inch air space. Then two sheets of $2 1/2$ pound lead were screwed on the sides of the channel and draped up over the spandrel. Where joints occurred in the sheet lead, closure was made by flat lock seams sealed with duct tape. The wall was finished on each side with $1/2$-inch thick gypsum board. All joints around the perimeter of the stubb wall were caulked to provide an airtight seal (Figure 7.54).

Figure 7.54 The mullion closures, $1 7/8$-inch thick sections that connect the dry walls with the mullions, are sound conditioned by the use of $2 1/2$ pounds per square foot ($5/128$-inch thick) sheet lead on each side of the closure. The closure ties the narrow mullion to the $4 7/8$-inch thick dry wall construction.

The machinery room walls were made of standard concrete block construction. To construct the sound barrier, firring strips were added to the inside wall surfaces of each room and the space between strips filled with acoustical wool. Gypsum board was then nailed over the strips, $2\frac{1}{2}$ pounds per square foot sheet lead ($5/128$-inch thick) was placed over the wall board, and joining edges of the sheet lead were folded into lock seams and sealed with duct tape. After the sheet lead was installed, a second layer of gypsum board was erected to finish the wall (Figure 7.55). This construction satisfactorily reduced the passage of noise originating from the rooms to acceptable sound levels.

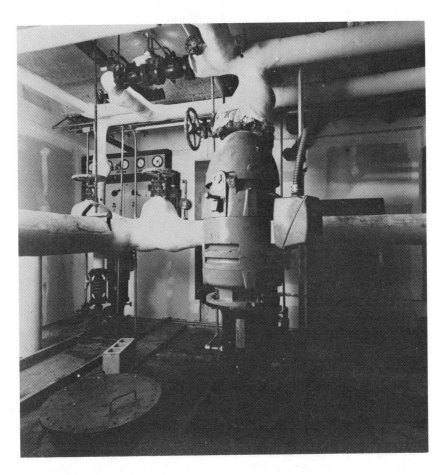

Figure 7.55 Sheet lead was also used on the walls of machinery rooms to reduce the noise generated by equipment to acceptable sound levels.

138 INDUSTRIAL NOISE CONTROL HANDBOOK

In the entire quieting procedure at the Hancock Center, over 12 tons of sheet lead were used to reduce noise to livable levels. According to the acoustical contractor, the sheet lead mullion sound barrier has proved so successful that to date the building management has had no complaints about noise (Figure 7.56).

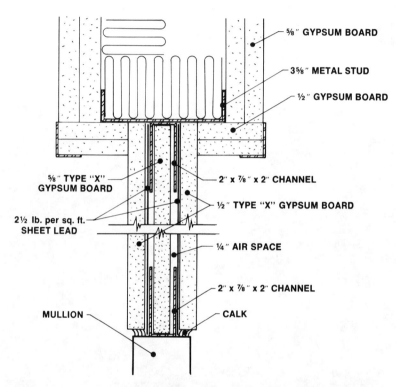

Figure 7.56 Line drawing giving construction details of mullion closures.

NOISE CONTROL PROBLEMS FOR PROVINCE OF QUEBEC JUSTICE BUILDING

If a judge presiding over court proceedings in Montreal's new Department of Justice Building for the Province of Quebec issues an order for "silence in the court," he can do so knowing that the noise he is hearing is coming from inside his courtroom rather than from outside sources. In fact, all working areas of the gleaming 17-story steel, glass and marble building— courtrooms, judges' offices, jury rooms, general offices, conference rooms— are designed so that noise from one area will not pass through walls or partitions to adjacent rooms. The result: the "Palais de Justice," as the building is known, probably offers more privacy than any building of its kind built.

Erwin Cleve, member of the Montreal firm of David & Boulva who designed the building, says the Palais de Justice is a classic example of the variety of noise techniques available to insure that working areas in a building are as quiet as possible. This, he said, is possible only with careful preselection of materials aimed at controlling those specific noise problems expected to be encountered after the building has been completed and in use. John Hillenbrand, project coordinator for the partitions contractor, Franz Patella, Inc., of Montreal, has reported that sheet lead solved most of these critical noise control problems. Included were sheet lead barriers above the tops of partitions and walls, as a component of soundproofing partitions and as noise barriers in sections of the building's heating and cooling system. In this latter instance, part of the building's heating and cooling system is enclosed in an open sheet metal housing extending around the interior perimeter of the building below the window level. Typical of those installed in high-rise buildings, it is a frequently overlooked channel for noise to travel the length and breadth of the building and into each room along the way. The problem was solved by installing a series of sheet lead baffles inside the sheet metal housing at the point where it intersects the partition line (Figures 7.57 to 7.59).

According to architect Cleve, prefabricated barriers of steel or gypsum board could not have been used because they could not have been made to fit around the various sizes of piping and ductwork making up the heating and cooling system. Instead, sheet lead was selected, not only because it blocks the path of noise but because it can be easily cut with knife or scissors and is simple to install. The sheet lead baffles were cut from a master template that yielded two unequal sections, which, when joined at the center by lock or standing seams, fitted around piping and vents extending through the system. Several hundred were required, each about $2^1/_2$ feet high and 15 to 18 inches deep. They were made of 2-pound lead. An extra $1^1/_2$ inches was provided around the edges of each baffle to allow for folding and adhering or mechanically fastening to the interior surfaces of the sheet metal housing. Finally, the lead was taped or wired at points of penetration for pipes and vents and all small cracks or openings were filled with an acoustical sealant. The combination of sheet lead baffle and adjacent office partition provided an installed sound transmission class (STC) of 45 for the total system.

Sheet lead also played a key role in providing barrier capabilities for a large number of other wall types, ceiling assemblies and partition supplements. Above inter-office partitions carried only to the ceiling, the plane

Figure 7.57 Close-up of heating system shows how sheet lead is installed at office partition wall line to keep noise from traveling between offices.

ENCLOSURES, SHIELDS AND BARRIERS—DESIGNING WITH LEAD 141

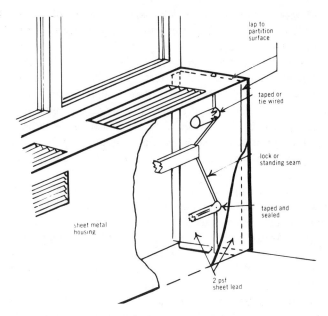

Figure 7.58 Diagram of the installation of sheet lead in the heating and cooling system. Conventional barrier materials could not be used because they could not be fit around the piping and ductwork.

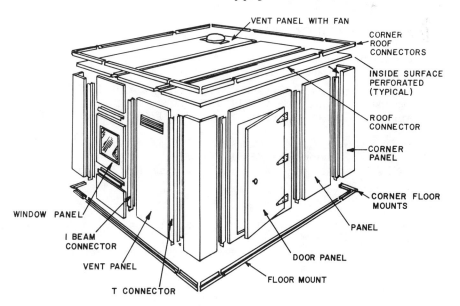

Figure 7.59 Typical enclosure with all its parts.

of the wall was continued to the slab above by means of a sheet lead curtain or plenum barrier. Depending upon requirements, either 1- or 2-pound sheet lead was used in these cases. The nearly 90 courtrooms in the Palais de Justice also required sheet lead plenum barriers, each about 2 feet high and a total of 80 feet long. Probably the most unusual of these courtrooms is the Palais de Justice Grande Court. A large dramatic room, it features stepped spectator seating, a court that is arranged around landscaped curved and rectangular tables rather than the traditional judge/jury box arrangement, and sweeping, curved walls. The sheet lead plenum barrier installed above the curved walls is nearly 3 feet high and about 175 feet long.

Some of the other applications for sheet lead included its use as one of the materials for acoustical wall panels for some of the offices near the building elevator core. Acoustical panels in this case varied in length but averaged 9 feet 6 inches in height. Altogether approximately 100,000 square feet of 1- or 2-pound lead was used for the Palais de Justice project according to Lead Industries Association (LIA). The Association points out that because sheet lead is one of the most efficient noise barriers of all common building materials, it is finding increased use for sound control applications to insure privacy for confidential business discussions and conferences, to soundproof executive offices and other areas in commercial or residential construction.

REFERENCES AND CREDITS

The following organizations were the principal sources of information, case histories, data and figures presented in this chapter.

1. Lead Industries Association, Inc., New York, New York 10017. Figures 7.1 to 7.31, 7.33 to 7.39, 7.45, 7.54 to 7.58.
2. Armstrong Cork Co., Lancaster, Pennsylvania 17604. Figures 7.40 to 7.44.
3. Singer Partitions, Chicago, Illinois 60611. Figures 7.46 to 7.50, 7.52.
4. ACON, Inc., Dayton, Ohio 45401. Figure 7.53.
5. Lord Allforce Acoustics, Inc., Erie, Pennsylvania 16512. Figure 7.59.

CHAPTER 8

NOISE REDUCTION WITH GLASS*

INTRODUCTION

Noise and its effects on man have received considerable attention in recent years and now rank among the most important problems in environmental control. Architects, home owners, and industrial manufacturers are all becoming more aware of the acoustical comfort and privacy problems associated with building construction. Additionally, studies have shown that although the normal ear can hear an amazing range of sound levels, prolonged exposure to high sound levels may lead to ear damage in the form of reduced hearing sensitivity.

Standard laboratory methods for testing the acoustical properties of materials have long been available. However, standard field methods for testing specific construction materials and their practical applications have been developed only recently. Unlike the controlled conditions of a laboratory, field testing takes place with conditions as they are. Such exploratory acoustical field tests have been conducted for PPG Industries. These field tests demonstrated that, in addition to selecting specific materials for sound reduction, it is equally important to consider partition materials and construction assembly details. Noise reduction measurements were all made in accordance with ASTME-336-67T, "Tentative Recommended Practice for the Measurement of Airborne Sound in Buildings."

Glass has been found to be an excellent sound reduction material. Inch for inch, it is better than most brick, tile or plaster and is about equal to medium-density concrete. Studies have also shown that as the thickness of the glass increases, the sound reduction, or sound transmission class (STC) rating also increases. However, increasing the interlayer thickness in laminated glass units does not increase STC ratings.

---

*By Pandit G. Patil, PPG Industries Inc., Glass Research Center, Pittsburgh, Pennsylvania 15238.

Further experiments have demonstrated that higher STC ratings can be achieved by the proper construction of double-glazed units. In addition, these units are superior to single or laminated glass in insulating against heat or cold.

SINGLE GLASS

General

Noise reduction can be achieved in three ways:
1. By reducing the sound level of the noise source;
2. By dissipating the sound near the receiver or the listener; and
3. By blocking the transmission path between the source and the receiver.

The first solution is the most effective, but also the most difficult to achieve in practice. The second solution is also most impractical. The third solution—blocking the transmission path—involves the placement of a relatively impervious partition somewhere between the noise source and the receiver.

Glass offers the potential for controlling noise and also establishing visual communication. The performance of a glass structure is primarily dependent upon the response characteristics it has to a spectrum of energy.

To understand the mechanism of sound transmission through partitions of any structure, one must study the single-glass wall because it forms the basic element of most structures. A modest average insulation of 35 dB can be obtained by using a single-glass wall weighing about 6 pounds per square foot. However, if an average insulation of 40 dB or more is required, the single-glass wall construction becomes decreasingly efficient, and the double-glass type of construction, which also provides good thermal insulation, should be used.

Theory

For airborne sound, the sound transmission coefficient (τ) of a panel is defined as the ratio of the transmitted sound energy to incident sound energy. The transmission loss or sound reduction index is given by the equation:

$$\text{TL} = 10 \log (1/\tau) \text{ dB} \tag{8.1}$$

The simplest theoretical treatment of single-glass partition involves the well-known mass law, which for sound waves at oblique incidence can be written as:

$$TL = 10 \; Log_{10} \left[1 + \frac{\omega \; M_s \cos \phi}{2 \; \zeta \; C} \right] \quad (8.2)$$

where: TL = the transmission loss
M_s = total surface density of the barrier (slugs/ft)
ϕ = the angle of incidence
ζ = density of air (slugs/ft^3)
ω = angular frequency (radians/sec)
C = velocity of sound in air (ft/sec)

If the incident sound wave impinges on the panel at normal incidence (when $\phi = 0$), then Equation 8.2 can be written as:

$$(TL)_o = 10 \; Log_{10} \left[1 + \left(\frac{\omega \; M_s}{2 \; \zeta \; C} \right)^2 \right] \quad (8.3)$$

where $(TL)_o$ is the transmission loss for normal incidence.

The quantity $(\omega M_s)/(2\zeta C)$ is generally large and, in practice, the sound waves impinge upon the barrier at a wide range of incidence angles. In this case, Equation 8.3 becomes Equation 8.4:

$$(TL) \; Field = 20 \; Log_{10} \frac{(\omega M_s)}{(2 \zeta C)} -5 \quad (8.4)$$

The value predicted by incidence field equation is the maximum that can be theoretically obtained with a single-glass barrier. When transmission loss is plotted against the frequency spectrum as a logarithmic scale, a straight line having a slope of 6 dB per octave results (Figure 8.1).

According to the mass law, a $7/32$-inch-thick glass wall has a potential average 38 dB but in fact only achieves 31 dB. This discrepancy is due to the influence of stiffness, which becomes significant at certain frequencies.

The mass law region is bounded by two important transition regions of stiffness-resonance and coincidence control. The low frequency transition boundary is generally defined by the stiffness response and the first few modes of natural frequency of the plate, whereas the upper boundary is limited by the coincidence frequency. Since these boundaries both tend to reduce the performance below mass law levels, it is imperative that they be thoroughly considered in the prediction of performance. These considerations become more pertinent and more complex for multiple glass and laminate structures.

Coincidence Frequency

The bending wave speed of a wall varies as the square root of the excitation frequency. At some combination of excitation frequency, and angle

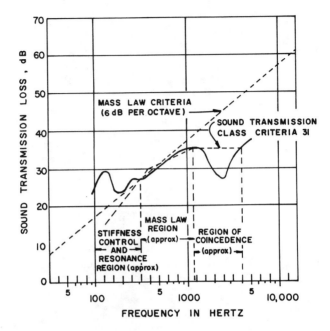

Figure 8.1 Transmission loss characteristics for $7/32$-inch-thick glass (3.0 lbs/ft^2). Laboratory sample $35^3/4 \times 83^3/4$ inches (tested in accordance with ASTM Designation E90-70).

of incidence, the trace wavelength of the incident sound wave will exactly coincide with the wavelength of the bending wave of the wall. The sound wave thus reinforces the bending wave, and a maximum amount of acoustical energy is transferred to the wall. At coincidence, the wall becomes quite transparent to sound and the transmission loss drops markedly.

Coincidence frequency is graphically explained below:

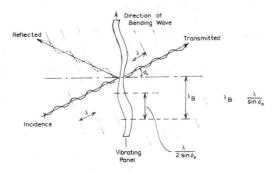

$$\lambda_B = \lambda$$

This condition of equal wavelengths in the glass panel and in the air is designated wave coincicence where:

λ_B = wavelength of the bending wave in the glass panel
λ = wavelength of the sound wave in the air

$$\lambda \sin\phi_o = \lambda_B$$

This means that the intensity of the transmitted wave approaches the intensity of the incidence wave. The frequency for which $\lambda = \lambda_B$ is called critical frequency. If the transmitted wave is as intense as the incidence wave, then the TL at that frequency and angle is very small.

Mathematically treated, coincidence frequency is expressed as:

$$f_\phi = \frac{\omega_\phi}{2\pi} = \frac{c^2}{2\pi \sin^2\phi} \sqrt{M/D}$$

where: C = velocity of sound through air
ϕ = angle of incidence
M = $\zeta_p h$
D = $Eh^3/12(1-\nu^2)$
ν = Poisson's ratio
ζ_p = mass density

$$f_\phi = \frac{c^2}{2\pi \sin^2\phi} \sqrt{\frac{\zeta_p 12(1-\nu^2)}{Eh^2}}$$

The experimental results for coincidence frequency are shown in Figure 8.2 as a function of thicknesses for the random incidence. It can be seen from the results that, as the thickness of glass plate increases, the coincidence frequency decreases.

Condition for Wave Coincidence

Wave coincidence occurs when

$$\sin \phi_o = \lambda/\lambda_B \qquad \sin \phi_o = C/C_B$$

C_B = velocity of propagation of the bending wave in the glass plate
C = velocity of propagation of the bending wave in air.

When the fixed frequency is assumed, the angle at which wave coincidence takes place is defined as the coincidence angle. When a fixed angle is assumed, the frequency at which wave coincidence takes place is defined as the coincidence frequency.

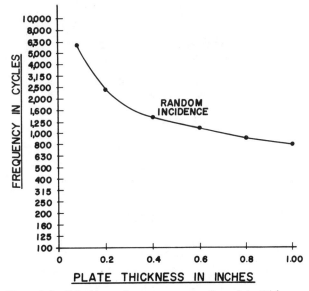

Figure 8.2 Coincidence frequency versus glass plate thickness.

Critical Frequency

The lowest frequency at which coincidence occurs is that frequency for which $\lambda_B = \lambda$ or $C_B = C$. Critical frequency is the lowest possible coincidence frequency and occurs for grazing incidence sound ($\phi_0 = 90°$). Mathematically treated:

$$f_c = \frac{C^2}{1.8hC'_L} \doteq \frac{C^2}{1.8h} \sqrt{p/E} \quad \text{cps}$$

where: h = thickness of the glass in ft for mixed English units

$$f_c = \frac{6.6 C^2}{hC'_L}$$

where: h = inches
c = ft/sec
p = lbm/ft^3
E = lbf/ft^2

These equations are valid only if $\lambda_B > 6h$

$$f \text{ coincidence} = \frac{C^2}{1.8 C'_L h \sin^2\phi_o}$$

$$f \text{ coincidence} = \frac{C^2}{1.8 h \sin^2\phi_o} \sqrt{\zeta p/E}$$

$$f \text{ coincidence} = \frac{1.16 C^2}{h \sin^2\phi_o} \sqrt{\zeta p/E} \quad \text{for mixed English units}$$

$$C'_L = \sqrt{E/\zeta p(1-\nu^2)} \doteq C_L$$

ν = Poisson's ratio 0.3 in most cases

Resonance

At the resonance frequencies of the glass wall, the transmission loss (TL) decreases sharply. For a hypothetical glass wall with zero damping, the TL would drop to zero at each resonance frequency. In actual walls, there is always some damping present so that the TL does not drop to zero. Also, because of damping, inertia, and the generally diminished power in the high frequency spectrum, significant reduction in TL usually occurs only at the lowest natural frequency. Since wall resonances may occur relatively close together in the frequency spectrum, the TL may be significantly below the mass law curve in a fairly broad frequency band.

Resonance frequency is expressed by the following mathematical relationship:

$$\omega_{res} = \sqrt{\frac{2.8 P_o}{d M_s \cos^2\phi} K_u}$$

where: K_u = correction factor
d = spacing of panels (ft)
M_s = mass of each panel (slugs/ft^2)
ϕ = angle of incidence
P_o = atmospheric pressure (lbs/ft^2)

Basic resonance is expressed as:

$$\omega_o = \sqrt{\frac{2.8 P_o}{dM_s} K_u} \qquad \psi\omega_{res} = \frac{\omega_o}{\cos\phi}$$

Figure 8.3 is a chart for determining resonance frequency, f_o, of two equal walls of combined surface density, M, separated by an air space of depth, d.

150　INDUSTRIAL NOISE CONTROL HANDBOOK

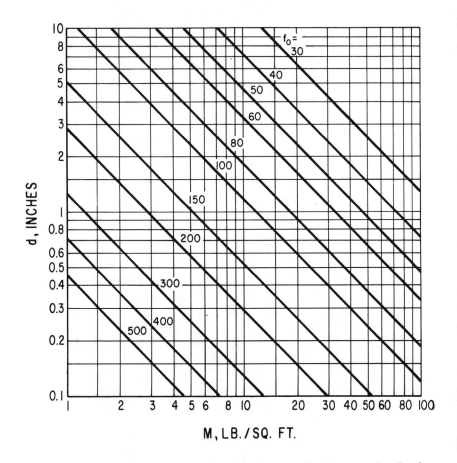

Figure 8.3 Chart for determining resonant frequency f_o of two equal walls of combined surface density M, separated by an air space of depth d.

Figure 8.4 is a schematic representation of transmission loss of a structure and its variation over the response spectrum of frequency as a function of the controlling parameters. With increasing frequency, these regions are stiffness controlled, resonance, mass controlled, coincidence, and supercoincidence.

It is apparent that the position of these regions in spectrum for a glass unit and the ability to control the positions are of great importance in establishing how a glass window shall perform as a transmission loss barrier component or system.

In the stiffness region, the performance is obviously totally undesirable. This being the case, it is standard procedure to establish the natural

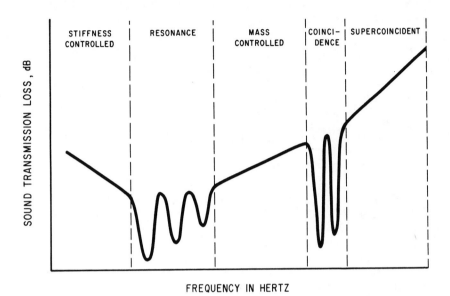

Figure 8.4 Schematic representation of TL of a structure and its variation over the response spectrum of frequency as a function of the controlling parameters.

frequency (first important mode) *below* the lowest frequency when transmission loss is required. This will assure that the window will perform well above the stiffness region throughout the spectrum of interest.

Sound Transmission Loss (STL)

The ratio of the sound energy incident upon one surface of a partition to the energy radiated from the opposite surface is called the sound transmission loss of the partition. Sound transmission loss is an inherent characteristic of a barrier and is essentially independent of the location of the barrier.

We do not actually "hear" the sound transmission loss, nor can we measure it directly. We can hear and measure the difference in sound pressure level between the spaces separated by a barrier, and we call this difference noise reduction. It includes the effect of the absorption present in the receiving room and the source room.

In test laboratories, the ASTM test method E90-70 is used. The test panel covers an opening between two rooms constructed with thick, massive walls that transmit much less energy than the test panel. Therefore, all of the transmission between rooms can be considered to take place through the test panel. The absorption in the receiving room is known.

A sound source is located in one room and the sound transmission loss is determined using:

$$NR = SPL_s - SPL_r$$

$$STL = NR + 10 \log_{10} S/A$$

where: NR = sound reduction
STL = sound transmission loss
SPL_s = sound pressure level in source room
SPL_r = sound pressure level in receiving room
S   = area of test panel
A   = total absorption in receiving room in units consistent with S

Experimental results showing STL as a function of frequency can be found in the *Acoustical Experimentation* section of this chapter.

Sound Transmission Class (STC)

To determine the sound transmission class (STC) of a test specimen, its sound transmission losses in a series of 16 test bands (determined in accordance with Recommended Practice E90-70) are compared with those of a reference contour having the form illustrated in Figure 8.5.

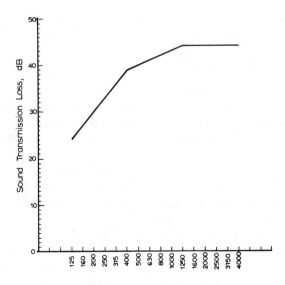

Figure 8.5 Typical STC contour.

If the transmission losses for the test specimen are plotted in a graph, the sound transmission class may be determined by comparison with a transparent overlay on which the STC contour is drawn. The STC contour is shifted vertically, relative to the test curve, until some of the measured TL values for the test specimen fall below those of the STC contour and the following conditions are fulfilled. The sum of the deficiencies (that is, the deviations below the contour) shall not be greater than 32 dB and the maximum deficiency at a single test point shall not exceed 8 dB. When the contour is adjusted to the highest value (in integral decibels) that meets the above requirements, the sound transmission class for the specimen is the TL value corresponding to the intersection of the contour and the 500-Hz ordinate.

The sound transmission classes (STC) for common materials are shown in Table 8.1

Table 8.1 Sound Transmission Class for Common Materials

Plate Glass Thickness (in.)	STC
$1/8$	28
$3/16$	31
$7/32$	31
$1/4$	31
$3/8$	32
$1/2$	35
$3/4$	37
1.0	38
$1/4$ inch steel plate	36
$3/4$ inch plywood	28
4 inch brick wall	41
6 inch concrete block wall	42
$1/2$ inch gypsum board on both sides of 2 x 4 studs	33
12 inch reinforced concrete wall	56
14 inch cavity wall (8 inch brick - 2 inch air - 4 inch brick)	65

Acoustical Experimentation

In a series of acoustical tests performed by PPG Industries at Riverbank Laboratories, eight thicknesses of single glass were tested for noise control. Figure 8.6 depicts the relationship of these eight glass thicknesses with regard to their sound transmission loss when subjected to noise frequencies ranging from 100 to 5000 Hz. The resulting sound transmission class ratings are shown in Table 8.2.

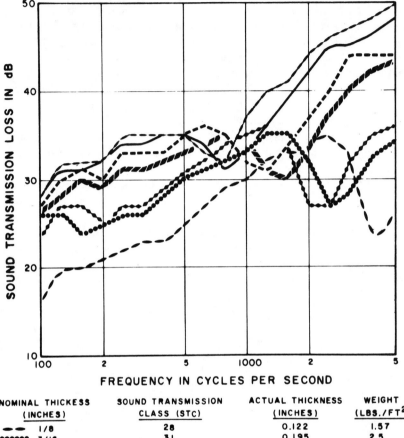

Figure 8.6 Single glass products and noise control (experimental results).

NOMINAL THICKNESS (INCHES)	SOUND TRANSMISSION CLASS (STC)	ACTUAL THICKNESS (INCHES)	WEIGHT (LBS./FT^2)
1/8	28	0.122	1.57
3/16	31	0.195	2.5
1/4	31	0.230	3.2
3/8	34	0.375	4.8
1/2	35	0.480	6.14
3/4	37	0.740	9.52
1	38	0.940	12.6

Generally, experimentation showed that, as the thickness of the glass increases, the sound transmission loss and resulting STC rating of the glass also increases.

Table 8.2 Sound Transmission Loss Through Single Glass[a]

Frequency (Hz)	STL $\frac{1}{8}''$	STL $\frac{3}{16}''$	STL $\frac{1}{4}''$	STL $\frac{3}{8}''$	STL $\frac{1}{2}''$	STL $\frac{3}{4}''$	STL 1.0''
100	16	26	23	26	27	28	27
125	20	26	27	28	30	31	32
160	20	24	27	30	31	31	32
200	21	25	25	29	30	32	32
250	22	26	27	31	33	34	35
315	23	26	27	31	33	34	35
400	23	28	29	32	33	35	35
500	25	30	31	33	35	35	35
630	27	21	32	34	36	34	33
800	29	32	34	35	35	31	32
1000	30	33	35	34	32	33	37
1250	32	35	36	31	31	36	40
1600	33	35	34	30	34	39	41
2000	33	32	27	33	37	42	44
2500	35	27	27	37	40	45	46
3150	33	28	32	40	44	45	47
4000	24	32	35	42	44	46	48
5000	26	34	36	43	44	48	50

Effects of Glazing

Studies were conducted to determine whether or not the type of mounting used for a single glass window had any effect on sound transmission loss. In the investigation, $\frac{3}{8}$-inch glass was mounted in four types of frames (see Figure 8.7):

1. neoprene gasket
2. metal frame with glass bedded in putty
3. wood frame with a wash leather strip around glass
4. concrete frame and putty.

Measurements were made according to ASTME 90-61T. At frequencies above 315 Hz, sound transmission loss values at a particular frequency were reproducible to within ± 1.5 dB. Below 315 Hz, variations of up to 8 dB made significant conclusions in this region impossible.

Figure 8.8 depicts the results of these studies. At all frequencies above 400 Hz, the window with the neoprene gasket gave a higher sound transmission loss than any of the other windows. There was no significant difference in the performance of the windows with wood and metal frames, but both of these were superior to the window in the concrete frame.

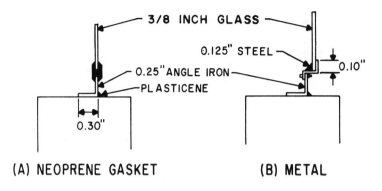
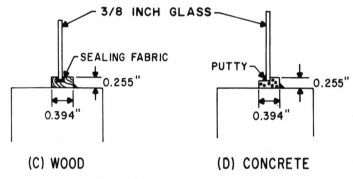

Figure 8.7 Window frames.

Maximum differences between the curves occurred at the coincidence frequency (1250 Hz), where the concrete frame gave a reading of 5 to 6 dB below that of the wood or metal frame and 10 dB below that of the neoprene gasket. The neoprene gasket configuration showed a sound transmission loss of 4 or 5 dB above that shown by the other frames at frequencies as low as 500 Hz.

DOUBLE GLASS

Theory

A double glass partition is defined as a partition in which points on opposite sides of the structure do not necessarily move the same way at the same time. It is commonly constructed by placing two single glass panes in series separated by an air gap. For the balanced partitions,

NOISE REDUCTION WITH GLASS 157

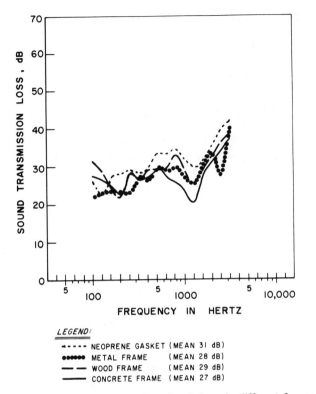

LEGEND:
- - - - NEOPRENE GASKET (MEAN 31 dB)
●●●●●● METAL FRAME   (MEAN 28 dB)
— — WOOD FRAME   (MEAN 29 dB)
——— CONCRETE FRAME (MEAN 27 dB)

Figure 8.8 Transmission loss of single windows in different frames.

the theoretical transmission loss of double partitions of infinite area at any angle of incidence is given by the following equation:

$$TL = 10 \log_{10} \left[1 + \left(\frac{\omega M}{2\zeta C}\right)^2 \cos^2\phi \left\{\cos\beta - \frac{1}{2}\left(\frac{\omega M}{2\zeta C}\right)\cos\phi \sin\beta\right\}^2 \right] \quad (8.5)$$

where: $\beta = \dfrac{\omega d \cos\phi}{C}$

d = spacing between the two glass panels

The transmission loss reaches a minimum value when the term within the bracket is zero, that is when:

$$\cos\beta = \frac{1}{2}\left(\frac{\omega M}{2\zeta C}\right)\cos\phi \quad (8.6)$$

At low frequencies, since d/C is small, Equation 8.6 can be written as:

$$f_\phi = \frac{1}{2\pi \cos\phi} \left(\frac{4\zeta C^2}{Md} \right)^{1/2} \tag{8.7}$$

This is an equation for resonance, and it has a particular value when $\phi = 0°$ and this value corresponds to normal incidence sound waves.

Figure 8.9 shows the typical experimental and theoretical curve for the balance unit. It also attempts to illustrate the type of transmission loss curve to be expected from a double glass partition. Different controlling regions are also shown.

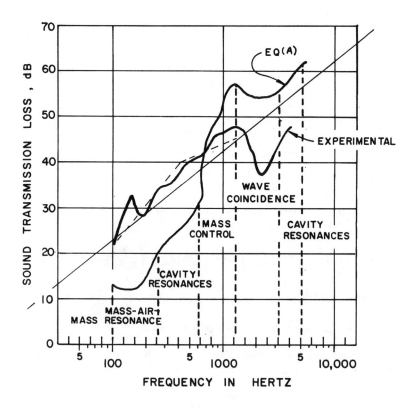

Figure 8.9 Transmission loss characteristics for double-glazed unit ($1/4$-in. glass, 2-in. air, $1/4$-in. glass).

The mathematical model for the theoretical transmission loss of unbalanced double glass units is given by:

$$TL = 10 \log_{10} \left\{ \frac{1}{2} \left[(\delta_1 - \delta_2)^2 + \eta_1^2 \right] \left[1 + \cos 2\phi_2 \right] + (\eta_2^2 + \delta_3^2) + \right.$$

$$(\eta_1 \delta_3 - \delta_1 \eta_2 + \delta_2 \eta_2) \sin 2\phi_2 +$$

$$\left. (\delta_1 \delta_3 - \delta_2 \delta_3 + \eta_1 \eta_2)(1 + \cos 2\phi_2) / 4\delta_3^2 \cos^2 \phi_2 \right\}$$

where: $\delta_1 = \zeta_1 C_1 \sec \sigma_1$ $\qquad \delta_2 = \zeta_2 C_2 \sec \sigma_2$

$\delta_3 = \zeta_3 C_3 \sec \sigma_3$ $\qquad \phi_2 = K_2 \beta_2$

$\eta_1 = \dfrac{m_1 \omega^2 - D_1 \kappa_1^4}{\omega}$ $\qquad \eta_2 = \dfrac{m_1 \omega^2 - D_2 \kappa_2^4}{\omega}$

$\beta_2 = L \cos \sigma$ $\qquad L = t_1/2 + d + t_2/2$

D_i = stiffness of ith wall
M_i = surface density of ith wall
K_i = radius of gyration of ith wall

Coincidence and Resonance Effects

In the case of double glass units, the effect of the coincidence frequency is the same as that for single glass units. The air space has little effect on coincidence frequency. Experiments show that the coincidence dip for double glass units occurs at the same frequency as that for the single glass units of the same glass thickness.

The problems associated with single glass, namely low-order wall resonances and coincidences transmission, occur in each panel of the double glass wall in just the same way. Mathematically:

$$f_{coincidence} = \dfrac{c^2}{2\pi h \sin^2 \phi} \sqrt{\zeta_p / E} \left(\sqrt{12(1-\nu^2)} \right)$$

Kurtze suggests that, in order to avoid a deep trough in the transmission loss curve due to the combined effects of low frequency resonance and coincidence, it is advisable to use panels with different bending wave velocities.

In addition to the problems associated with single glass units (low-frequency resonance and coincidence effects), other types of resonance influence the sound transmission loss of double glass units. Figure 8.10

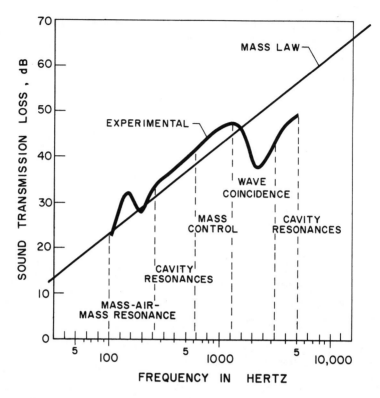

Figure 8.10 Sound transmission loss characteristics for matched double-glazed unit ($1/4$-in. glass, 2-in. air, $1/4$-in. glass).

illustrates their behavior. The resonance does not always occur in separate regions.

For double glass units, the sound transmission loss depends on several factors including frequency, weight, method of edge restraint, damping and stiffness. The most important variable of all is edge **restraint**. It has been proven experimentally that two different edge restraints with the same glass combination give different sound transmission class ratings. Experimental results are given in the section entitled *Effects of Glazing*.

Design of Double Glass Units

The following three parameters must be considered when designing double glass units: air space, glass thickness and damping.

Air Space

Generally, increasing the air space between two glass panels results in increased sound transmission loss. However, studies show that maximum sound insulation is reached with the optimum spacing of four inches. An air space wider than four inches acts as a convective space for vibrating waves, thereby reducing performance. Experimental results are given in the *Acoustical Experimentation* section.

Glass Thickness

A combination of glass thicknesses can improve the performance of the unit by overcoming the resonance and coincidence effects. The $1/4$-inch and $3/8$-inch glass combination, along with restrictions on the air space, gives the best performing unit. Experimental results for 2-inch air spaced units with different glass combinations are given in the *Acoustical Experimentation* section.

Damping

Damping is extremely important to the performance of the unit. The experimental results show that a $3/8$-inch and a $1/4$-inch glass panel, separated by a 2-inch air space with proper damping in the edge attachment, has an STC rating of 45. With little or no damping, the sample has an STC rating of 42.

Acoustical Experimentation

Four experimental tests were performed using two pieces of $1/4$-inch glass separated by various air spaces (1 inch, 2 inches, 3 inches and 4 inches). These studies were conducted to determine the relationship between the size of the air space and the sound transmission loss rating of the window. The studies showed that the sound transmission loss increased as the air space increased. For example, the sound transmission class (STC) rating rose from 39 for the $1/4$-inch glass, 1-inch air space, $1/4$-inch glass configuration. Figure 8.11 graphically displays the experimental results, and they are also compared in Table 8.3.

Three additional studies were performed in an effort to determine the effects of glass thickness on the sound transmission loss rating. In each case, there was an air space of two inches, and the thicknesses of the glass panels were varied. Three configurations were used: $1/4$-inch glass, 2-inch air space, $3/16$-inch glass; $1/4$-inch glass, 2-inch air space, $3/8$-inch glass; and $1/4$-inch glass, 2-inch air space, $1/2$-inch glass.

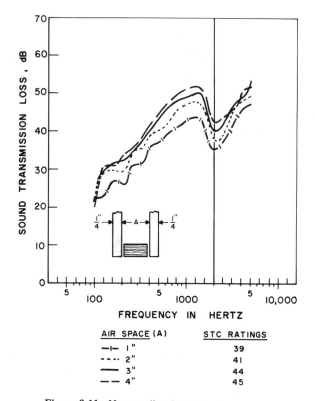

Figure 8.11 No coupling between glass panes.

Experimental results are shown graphically in Figure 8.12. In general, the $1/4$-inch glass, 2-inch air space, $3/8$-inch glass configuration was the best performing unit. They are also compared in Table 8.4.

Effects of Glazing

Experiments were conducted using double glass units with different edge attachment in an effort to determine whether or not the type of mounting had any effect on the sound transmission loss. Results using a $1/4$-inch glass, 2-inch air space, $3/8$-inch glass configuration demonstrated that the sound transmission loss was greater with the resilient edge attachment than with the stiff edge attachment (Figure 8.13).

In other studies, a 1-inch Twindow® unit ($3/8$-inch glass, $7/16$-inch air space, $3/16$-inch glass) with resilient edge attachments was mounted in both wood and aluminum frames. Experimental results showed that the

Table 8.3 Sound Transmission Loss Through Double Glass Units with Varying Air Space

Frequency (Hz)	Sound Transmission Loss			
	$1/4''$ glass $1''$ air space	$2''$ air space	$3''$ air space	$4''$ air space
100	23	22	21	20
125	23	29	30	31
160	27	30	32	31
200	26	29	32	33
250	31	34	35	35
315	31	35	37	38
400	36	39	41	42
500	37	40	44	45
630	40	42	47	48
800	41	46	48	50
1000	43	47	49	51
1250	44	48	50	52
1600	41	45	46	48
2000	35	37	40	42
2500	37	39	42	44
3150	42	45	47	46
4000	46	48	49	51
5000	47	49	53	52

unit mounted in the wood frame had a superior sound transmission class rating. The experimental results are shown in Figure 8.14.

Other 1-inch Twindow units ($3/8$-inch glass, $7/16$-inch air space, $3/16$-inch glass) with stiff edge attachments were also mounted in both wood and aluminum frames. Experimental results showed little difference in STC between the wood frame and aluminum frame units. These experimental results are shown in Figure 8.15.

MULTIPLE GLAZING

Windows with more than two glass panes are usually restricted to special situations, such as observation windows in studios. Therefore, because of their limited application, they will be discussed in less detail than the single or double windows. Theoretical studies have shown that a multiple unit readily transmits energy below a limiting frequency dependent on the mass of each panel and the distance between the glass panels. Above this frequency, the slope of sound transmission loss as of function of frequency becomes greater as the number of glass panels increases.

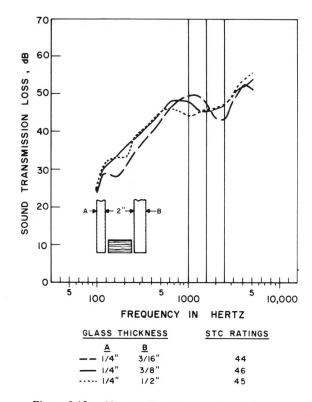

Figure 8.12 No coupling between glass panes.

Brandt measured the insulation of double and triple windows (2, 3 and 4 mm glass and air space from 32 to 160 mm). In each case, the air spaces of the triple window were equal. Although the average insulation of the triple windows was between 0 and 2 dB greater than that of the corresponding double window, the majority of triple windows had a greater sound transmission loss than the double windows at frequencies above 500 Hz, and a lower sound transmission loss than the double windows at frequencies lower than 500 Hz.

If a triple window is formed by adding a panel of glass between and parallel to the panels of an existing double glass window forming two equally wide air spaces, the mean sound transmission loss of the triple window is almost equal to that of the double window. Some improvement may be made if the air spaces are of unequal width, but an undivided air space gives better insulation (greater sound transmission loss) at low frequencies.

Table 8.4 Sound Transmission Loss Through Double Glass Units with Varying Glass Thicknesses

Frequency (Hz)	Sound Transmission Loss		
	1/4" glass 2" air space 3/16" glass	1/4" glass 2" air space 3/8" glass	1/4" glass 2" air space 1/2" glass
100	25	25	33
125	31	32	36
160	33	33	29
200	36	33	34
250	38	38	38
315	40	40	36
400	43	43	42
500	45	45	46
630	48	46	50
800	48	45	51
1000	48	44	50
1250	46	45	50
1600	45	46	51
2000	46	46	46
2500	47	47	48
3150	50	51	51
4000	52	54	54
5000	54	56	55

LAMINATED GLASS

Theory

The response of a laminate to a random sound field is determined by the material and geometrical parameters of the laminate. As shown in Figure 8.16 the response spectrum is made up of several regions. These regions, with increasing frequency, are stiffness controlled, resonance, mass controlled, coincidence and supercoincidence. There are also transition regions that can be important considerations.

The location of these regions in the spectrum and the band width of each region is determined by the materials used in the laminate, and the cross-sectional and areal geometry. It is apparent that the position of these regions in the spectrum for a laminate and the ability to control the positions are of great importance in establishing how a laminate will perform as a transmission loss barrier component or system.

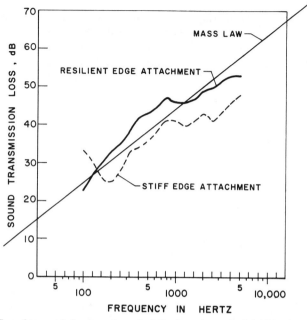

Figure 8.13 Sound transmission loss characteristics for mismatched double-glazed units with different edge attachments ($1/4''$ glass, $2''$ air, $3/8''$ glass).

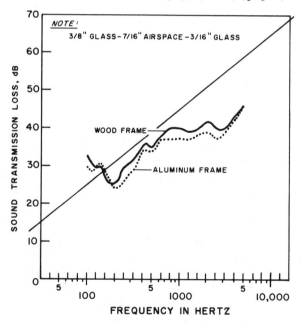

Figure 8.14 Effects of glazing on double glass windows.

NOISE REDUCTION WITH GLASS   167

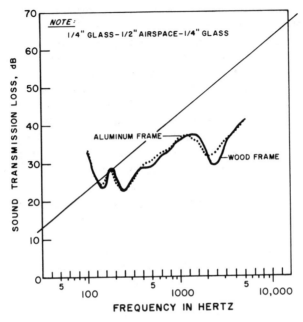

Figure 8.15 Effects of glazing on double glass windows.

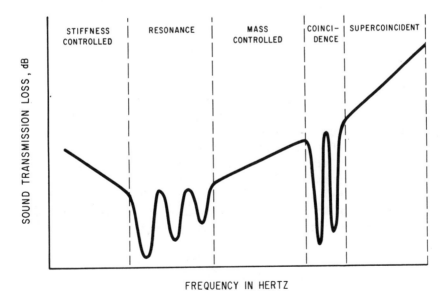

Figure 8.16 Schematic representation of TL of a structure and its variation over the response spectrum of frequency as a function of the controlling parameters.

In the stiffness region, the performance is obviously totally undesirable. Because of this, it is standard procedure to establish the natural frequency (first important mode) below the lowest frequency where transmission loss is required. This will assure that the laminate will perform well above the stiffness region throughout the spectrum of interest.

The resonance region performance is the most complex of the regions under consideration, and the response characteristics are most important. With the exception of the transition region between mass law and coincidence, the resonance region is the only range in which damping can be effectively used to control noise. For this to be achieved, the laminated structure must be designed to extract more energy from the principle acoustic radiating modes than inherent types of damping (*i.e.*, joint friction, air damping and acoustic radiation damping by pressure cancellation). Therefore it is important to identify the principal radiating modes for a laminate by examining the resonant frequencies in response to the airborne sound field and structure-borne sound energy. It is evident that in a random sound field the modes of response of the laminate will be many, and it is necessary to select the dominant modes (*i.e.*, high amplitude response) and apply the damping to these modes. This, of course, is admitting that the damping of a constrained laminate configuration is frequency sensitive, which means that the damping can only be applied to a region of limited frequency.

Damping of fundamental modes to reduce noise is normally a last resort. Preference should always be given to shifting the fundamental to a lower frequency, out of the spectrum of interest if possible.

The mass law region, in a practical sense, is the most desirable area in which to place laminate response. In this region the transmission loss increases at the rate of 6 dB per octave of frequency. With the exception of the coincidence region, in which it is impractical to achieve any low frequency performance, the mass law region represents the optimum area in which to position the laminate response. This positioning can be accomplished by variation of the material and geometrical parameters.

Between the mass law control and coincidence regions there is a resonance effect of higher order modes that again causes a sharp drop in the transmission loss properties of the laminate. If this notch falls into the spectrum of interest, then damping in this frequency range becomes important.

The coincidence region is important when the sound pressure level of a field is rich in high-frequency content. The transmission loss in this region increases at the rate of 9 dB per octave change in frequency as a nominal figure under random incidence (*i.e.*, 12 dB per octave is theoretically obtainable under normal incidence but has not been practically realized). As mentioned previously, the coincidence region cannot be shifted to lower frequencies without introducing impractical configurations, particularly when restricted to glass skin laminates with polymeric, viscoelastic cores. It is important to note that a laminate can be effectively used to perform in a mass law-transition-coincidence region with a high degree of efficiency when the sound field is skewed toward high frequency response.

The Natural Frequency of a Laminate

The natural frequency of a plate is dependent upon its mass, bending stiffness, boundary conditions, and size and shape effects. The relationship given by Ritz for homogeneous isotropic square plates is as follows:

$$\omega_n = \frac{\alpha}{\ell^2} \sqrt{B/M}$$

where: ℓ = length of one side of the plate
α = a constant depending upon the plate geometry and boundary conditions (often called the mode factor).

The application of the Ritz equation to laminated structures is complicated by the fact that the bending stiffness (B) of a three-ply laminate is a function of the excitation frequency (ω). Using these relationships (keeping in mind the limitations imposed in the simplified form), the application of the Ritz equation to laminate performance for most cases becomes routine.

Designing a laminate to perform as an effective transmission loss barrier in the resonance region as a function of damping is a complex task requiring careful consideration of many parameters and the judicious selection of materials. As previously stated, it is best to avoid the resonance region whenever possible, and the objective, at least at moderate frequencies, should be mass law performance. At higher frequencies, of course, this is more realistically achieved. It is also necessary to retain cognizance of the several factors that can override the performance of the most rigorously conceived laminate such as structure-borne noise of high amplitude (at low frequency), temperature displacement of peak damping, and the damping of the laminate due to its configuration. versus that damping induced by structural edge restraint of the laminate (such as friction or air damping).

Other complexities that enter into the picture are strain sensitivity of certain polymers, the practical experimental problem of making high frequency (as far as the audible spectrum is concerned $\approx$ 22,000 cps) measurements of loss factor and shear modulus for the core materials and, of course, all of the practical material problems such as structural bonding.

Experimental Results

The laminated wall represents an attractive method of combining all the desirable characteristics for TL. It is possible to design a laminated wall with high static stiffness, low dynamic stiffness, high damping, and a critical frequency above necessary limits. This type of wall utilizes the fact that structures with stiff skins separated by an incompressible spacer have bending stiffness characteristics that vary inversely with frequency squared. If the spacer is made of a material with low shear stiffness (*e.g.,* rubber), shear waves can be made to propagate in the core. Using these principles, it is possible to design a limp, well-dampened wall that also fulfills the necessary static requirements. A well-designed laminated wall may approach TL characteristics as predicted by the mass law.

Figure 8.17 shows some experimental results for PPG's laminated glass products. Two pieces of $1/8$-inch glass were laminated with various thicknesses of interlayer. The STC for all three cases was 34, but the performance curves differed markedly.

Effects of Temperature on TL

Figure 8.18 shows the test results for 1-inch acoustical Twindow units at various temperatures. At the normal condition, with the source room at 71°F, 45% RH, and the receiving room at 72°F, 89% RH, the STC was 38. When the source room was at 110°F, 25% RH, and the receiving room at 71°F, 91% RH, the sound transmission loss curve improved 1 or 2 dB at the low-frequency end, and decreased about 1 dB at the high-frequency end. However, this increase or decrease in the sound transmission loss is not sufficient to change the STC rating of the unit. When the source room was at 32°F, 54% RH, and the receiving room at 71°F, 85% RH, the sound transmission loss curve improved at the low and high frequencies by about 1 or 2 dB. There is also some improvement in the middle frequency range. The effect of all of this is to increase an STC rating by 1 dB.

It can be seen from Figure 8.19 that temperature has a great effect on the sound transmission loss of a laminate. The example shown in Figure 8.19 is $1/8$-inch glass, 0.045-inch interlayer, $1/8$-inch glass laminated when the source room was at 71°F, 45% RH, and the receiving room was at 71°F, 89% RH (standard conditions). The sound transmission class was

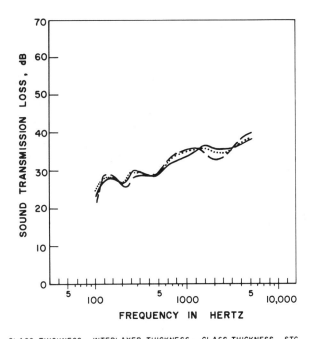

GLASS THICKNESS (INCHES)	INTERLAYER THICKNESS (INCHES)	GLASS THICKNESS (INCHES)	STC
—— 1/8	0.030	1/8	34
······ 1/8	0.045	1/8	34
– – 1/8	0.015	1/8	34

Figure 8.17 PPG's laminated glass.

34. Notice the high and low ends on the spectrum curve. At the low end there are two resonances. At the high end there is minimal drop due to coincidence. When the source room was at 20°F, 54% RH, and the receiving room at 71°F, 85% RH, the sound transmission class dropped sharply to 31. The great drop in sound transmission loss occurred at the coincidence frequency. There is also some drop at the low-frequency region. When the source room was at 110°F, 25% RH, and the receiving room at 71°F, 91% RH, the sound transmission loss dropped sharply at the low frequency region; at the middle- and high-frequency regions it increased considerably. The total effect of this was to increase the STC rating to 38.

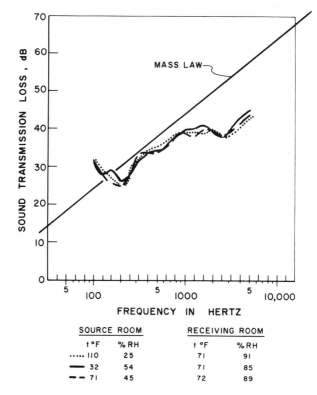

Figure 8.18 Effect of temperature on sound transmission loss for double-glazed window.

PPG ACOUSTIC WINDOW

PPG Industries presently produces a 1-inch Acoustic Twindow unit with an STC rating of 38 compared to the standard rating of 32 for a standard 1-inch Twindow unit. Table 8.5 provides the sound transmission class (STC) data comparing the Acoustic Twindow unit with other windows. Figure 8.20 compares the sound transmission loss of Acoustic Twindow and Standard Twindow as a function of frequency.

In Acoustic Twindow units, each panel of glass is a different thickness. The panels are separated by a $7/16$-inch air space hermetically sealed with a special acoustically dampened perimeter section.

Acoustic Twindow units are recommended when transmitted noise must be reduced but visual communication is desired. They may be used

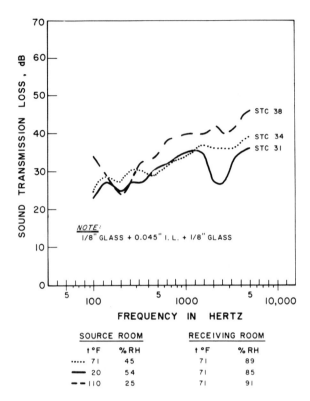

Figure 8.19 Effect of temperature on sound transmission loss for laminated glass.

Table 8.5

Product	STC	Approx. Weight lb/sq ft
Acoustic Twindow		
3/16 inch glass, 7/16 inch air space 3/8 inch glass	38	7.8
Standard Twindow		
1/2 inch air space	32	6.0
9/32 inch laminated glass		
0.045 inch vinyl	34	3.5
1/2-inch float	34	6.0
1/4-inch float	31	3.2

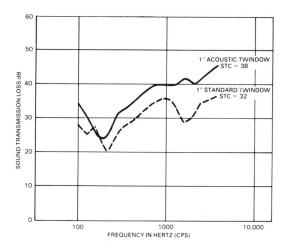

Figure 8.20 Sound transmission loss (ASTM E90-70) Acoustic Twindow.

outdoors or indoors, in walls and partitions, in windows and doors. The Acoustic Twindow unit offers the brilliant clarity of float glass or a choice of Solarbronze or Solargray tinted glasses.

These units are available in clear glass combinations to a maximum area of 70 square feet. Tinted glass combinations are available to a maximum area of 50 square feet (available in larger areas depending upon special processing required). Minimum size is 16 x 24 inches; thickness glass-to-glass is 1-inch, or $1^1/_{16}$-inches in overall thicknesses at the enclosing metal edge (subject to normal manufacturing tolerances). The average weight of the 1-inch unit ($^3/_{16}$-inch and $^3/_8$-inch glass) is approximately 7.8 pounds per square foot.

CHAPTER 9

ADDITIONAL SOUND CONTROL MATERIALS

In addition to lead and lead-loaded products, there are many other materials used for sound attenuation and vibration control. In this chapter we shall examine these other materials and some of their applications.

Materials with high sound absorption qualities generally have soft porous surfaces. When sound waves come in contact with these absorptive surfaces, air travels in and out of the pores in the material because of pressure changes produced by the sound. Frictional forces that result from this action convert the sound energy into heat, even though the initial amount of energy is quite small. Every time sound waves travel through the material at each reflection, energy is dissipated, and there is a resulting reduction in the reverberant sound level of the particular enclosure. The use of acoustical material can cut down sound levels in an entire plant as well as limited areas.

ACOUSTICAL FOAMS[1]

Many different types of foams are used widely by industry for sound and vibration isolation. Most industrial foams are available in pore size from 10 ppi to 100 ppi (pores per linear inch). Textures are generally coarse and abrasive in the 10 ppi grades to soft and downy in the 100 ppi grades. Acoustical foams that have excellent high and low temperature features can withstand temperatures as high as 250°F, so they can be sterilized with boiling water or steam.

One of the most important factors determining the behavior of a foam sound absorber is *its resistance to air flow*. In flexible urethane foams, this concept is called "permeability." Permeability measurements are made by first measuring air flow resistance of the foam and then inverting. For example, a completely closed cell foam would totally resist air flow and have a zero permeability. As a more open cell or reticulated foam is achieved, the measurement of permeability rises to 30%, 50%, 70% and so on.

A common misconception concerning flexible urethane foams is the belief that two foams that look alike (have the same thickness, cell size and density) will provide equal sound absorption. In reality, these two foams may have permeabilities differing by several orders of magnitude and have widely different acoustical properties. To understand this one must understand how sound or noise becomes absorbed. In principle, noise contacts the foam structure in the form of sound pressure waves. The pressure wave within the foam structure is converted to heat energy and is dissipated. The ability of the foam to absorb sound is determined primarily by its permeability and its thickness. Thickness is easily seen and measured, but permeability is neither easily measured nor visually compared.

A very practical, on-the-spot-method of determining whether a foam has permeability is to put it against your lips and blow on it. If it totally resists the pressure you put on it, the foam is not permeable. If you can blow through it, it has a degree of permeability. To understand the reason for the enormous variations of permeability encountered in flexible urethane foams, it is necessary to know something about foam geometry. Urethane foams are formed by mixing reactants that simultaneously polymerize to yield a polymer and generate a gas, which causes the reacting mass to expand into a foam.

Initially these gas nuclei are very small (the number of gas nuclei generated per unit volume determines cell size). However, as the reaction proceeds the gas bubbles grow larger until, at the point where the porosity of the expanding mass has reached a level of about 72% (density of about 20 PCF), the gas bubbles become distorted into polyhedrons and the contact points develop into planes. At the instant of completion of expansion the fully expanded low density foam consists of nested polyhedrons (ideally these are dodecahedrons having 12 facing planes). Each common face between the polyhedrons contains a thin polymer film or membrane. At this point, the result is a closed-cell foam, completely pneumatic and having zero permeability.

To produce an open-cell foam it is necessary to design the formulation and processing conditions so that at the exact point the polymer mass is expanded and set into a foam of the desired density, some of the films or membranes separating the cells collapse and establish communication between the cells. Assuming the cells are dodecahedrons, it is necessary for one-sixth of the membranes to open to achieve an open cell foam.

Thus urethane foams have a strand and membrane configuration, the strands defining the edges of the nested polyhedrons and the membranes the planes or faces (often called pores) of the polyhedrons. The permeability of a foam is proportional to the number of ruptured membranes. The membranes comprise only a very small portion of the polymer mass, about 1-2%. In the foam industry, the term "level of reticulation" is often used to indicate the percentage of the membranes that are missing. Foam containing no membranes is said to be 100% reticulated.

The number of membranes present in a foam is, of course, the major factor in determining its permeability. Expressing permeability as its inverse, flow resistance, fully reticulated foams have a flow resistivity ranging from about 20 Rayls per inch for a very fine cell foam to about 1 Rayl per inch for a very large cell foam. However, a foam of any cell size which has most of its membranes intact, will exhibit a very high flow resistivity, often greater than 1000 Rayls per inch.

The conventional bun stock process for urethane foam manufacture presents the foamer with some formidable obstacles. The bun or loaf of foam, perhaps 36 inches thick, emerging from the machine and destined to be cut into sheets has an inherent variation in permeability from top to bottom. Maintaining average permeability involves cutting and testing sections, so rapid feedback for process control is difficult, if not impossible.

The optimum permeability for an absorber depends upon its thickness. As a general rule the thinner the sheet, the higher its flow resistivity to achieve optimum performance. Again, the thin-sheet casting process is advantageous since the permeability of the single sheet being produced can be optimized for its thickness. However, the bun stock manufacturer is faced with the choice of compromising on an average permeability or producing buns having different permeabilities for each sheet thickness. The second choice leads the bun stock manufacturer to the additional serious problem of controlling inventory.

The ability to control permeability in foam casting and thus achieve predictable acoustical results has many advantages for an engineer concerned with noise control. Principally, he can now select a foam absorber having good performance in any desired frequency range. These factors can be written into specifications with the assurance that the supplier can deliver a consistently uniform foam absorber for each application.

FOAM WITH POLYMER FILM SURFACE[2]

A foam consisting of an acoustic foam for absorbing sound at peak efficiencies combined with a polymer film surface is now being marketed.

The polymer film surface resists scuffing and abrasion, is self-repairing against punctures and has no tendency to tear.

The material is specifically designed for areas demanding a high degree of sound absorption in a visible foam surface. Sheets 24 x 54 inches and $1/2$, $3/4$ or 1 inch thick come in a hexagonal pattern that can be folded, bent, or worked into complicated configurations without degrading their acoustical efficiency. Its temperature range is from -40°F to 250°F, and it is resistant to petroleum and alkalis. This product is produced by the Soundcoat Co., Inc., New York, New York, and its sound absorption qualities are illustrated in Figure 9.1. Figure 9.2 illustrates a typical application for this product.

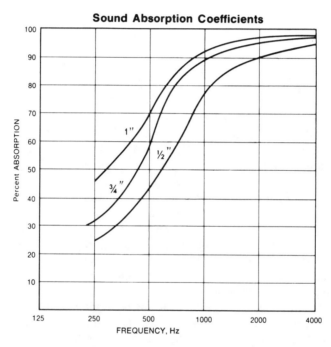

Figure 9.1 Acoustical performance of a foam with a polymer surface.[2]

ACOUSTICAL PANEL[3]

An acoustical panel is a composite of a 5 mil white polyester film chemically bonded to a foam structure that utilizes controlled permeability for optimum acoustical performance. The product also utilizes a pressure-sensitive adhesive backing for ease of application. The film facing,

ADDITIONAL SOUND CONTROL MATERIALS 179

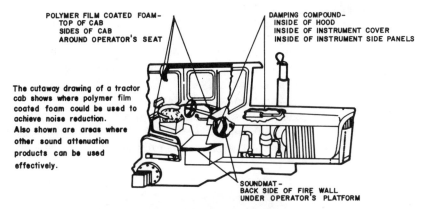

Figure 9.2 Application for polymer-coated foam.

combined with the proper air flow resistance of foam, yields excellent low frequency absorption performance. The facing, while being attractive, is durable and well suited for use in environments in which it would be exposed to rough handling, grease, moisture, solvents and other contaminants.

The acoustical panel has good low-frequency performance; it is produced by Specialty Composites Corp. Figure 9.3 illustrates its acoustical qualities at various frequencies.

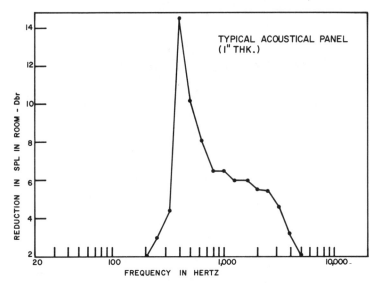

Figure 9.3 Sound qualities of an acoustical panel.

FLEXIBLE POLYURETHANE FOAM[3]

Figure 9.4 illustrates samples of Ferro Coustifoam® 3-D sound absorption foam. It is a flexible polyurethane foam with a random coefficient of absorption equal to 2 lb/cu ft of fibrous glass, the most efficient noise absorber known over a broad frequency spectrum. Available in standard thicknesses of $1/2$ and 1 inch and with pressure-sensitive self-adhesive backing for easier installation, it is nontoxic and nonirritating to handle. Requiring little mechanical support because of high tear and shear strengths, it can be free-hung on wires or attached to vertical surfaces with adhesives.

Figure 9.4 A flexible polyurethane foam available on the market.

This foam is a sound absorption material used for middle and high-frequency applications in which an attractive appearance is desired, such as hard walls and vehicle headliners. It is also effective as absorbing baffles. Because it can be formed to compound curves with a minimum wrinkling, it is used for vehicle headliners. It can also be used as absorbing liners for appliances, truck cabs, golf carts, snowmobiles, and office equipment.

For low-frequency applications, the efficiency of this product can be improved by spacing it away from a hard surface. It has an expected service life of 10 years under normal, in-plant, or vehicle conditions. It will meet random incident absorption coefficient, aging, durability and flammability requirements. Figure 9.5 and Table 9.1 give data on the absorption coefficients of this material.

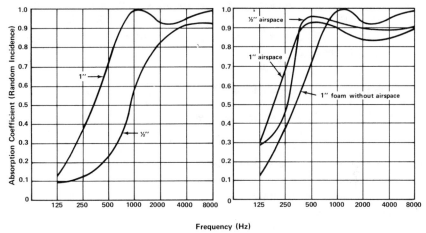

Figure 9.5 Acoustical qualities of polyurethane foam.[3]

Table 9.1 Absorption Coefficients of Polyurethane Foam

Thickness (inches)	Frequency (Hz)							NRC[a]
	125	250	500	1000	2000	4000	8000	
1/2	0.90	0.11	0.22	0.60	0.88	0.94	0.90	0.45
1	0.13	0.22	0.68	1.00	0.92	0.97	0.99	0.71

[a]Noise Reduction Coefficient (NRC) is the average of absorption coefficients at 250, 500, 1000, 2000 Hz. Coustifoam can be "tuned" to provide maximum NRC by varying its spacing from the mounting surface. The Ferro technical representative can give specific recommendations.

VARIOUS USES FOR POLYURETHANE FOAM[4]

Flexible polyurethane foam is a highly versatile material that can solve a wide variety of acoustical problems. The number of newly acclaimed uses is constantly growing, based on different specific application requirements and unusual foam structure designs that successfully meet them. In the increasingly important area of noise reduction, polyurethane foam has proven to be practical. There is no secret ingredient involved in the sound-absorption successes of polyurethane foam. Actually, what is most important about it is what is missing.

Through a special reticulation process, developed and patented by Scott Paper Company's Foam Division, all the natural membranes between foam cell strands are removed. The remaining skeletal strand structure of the fully reticulated "open-pore" polyurethane foam is then better able to absorb noise, causing a loss of sound energy by weakening sound waves through reflection.

IMPORTANT VARIABLES IN SELECTION

Scott acoustical laboratories, using an impedance tube and following the test procedures outlined in ASTM-C384, investigated the effects of many variables on the ability of flexible polyurethane foam to absorb unwanted sound. These variables included permeability, pore size, thickness, density, stiffness and surface treatment.

The effect of permeability on the overall absorption capability of foam proved interesting. In general, completely closed-cell foam shows minimal capability to absorb sound energy, regardless of thickness. Some open-cell foams have excellent absorptiveness in 1-inch to 2-inch thicknesses. Fully reticulated open-pore foam in thicknesses of 2 inches and greater has outstanding broad-band absorption capabilities.

In testing the effect of pore size, foams ranging from 20 to 93 ppi (pores per lineal inch) were used. These tests showed that, with all other variables constant, the smaller the pore—or the greater the number of pores per lineal inch—the greater the absorption capability. Also, the thicker the piece of a particular foam, the greater the noise reduction. For example, the sound energy absorbed was measured by various thicknesses of an 80 ppi, fully reticulated open-pore foam. Some of the results are: 6-inch foam absorbed nearly 100% of the noise from 250 to 6000 Hertz; 4-inch foam absorbed 97% noise from 1000 to 6000 Hz, 93% at 500 Hz, and 60% at 250 Hz; 2-inch foam absorbed 95% at 4000 to 6000 Hz, 52% at 500 Hz (see Figure 9.6).

To determine the effects of density and stiffness, reticulated foams ranging from 1 to 6 pounds per cubic foot were first tested, and then tested again after being rigidized by means of plastic coatings and metalizing. Resultant variations in absorption capabilities were negligible in all cases.

In certain noise control applications it is necessary to protect the foam from its environment. This can be accomplished by laminating a film to the foam, making it more resistant to abrasion and impervious to liquid penetration. However, complete lamination of film surface to foam gives good absorptive characteristics in only a narrow frequency range. It has been found that a partial lamination considerably improves the sound absorption characteristics of the package.

ADDITIONAL SOUND CONTROL MATERIALS 183

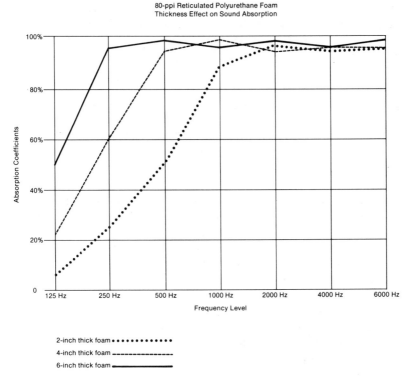

Figure 9.6 In testing the effects of many variables on the ability of polyurethane foam to absorb unwanted sound, Scott acoustical lab technicians used an impedence tube and followed the test procedure in ASTM-C384. It was found that the fine-pore, fully reticulated foams are excellent sound absorbers in thicknesses of 2 inches and above. As an example, the sound energy levels absorbed by various thicknesses of an 80 ppi (pores per lineal inch) fully reticulated foam were measured. The results of the 2-, 4- and 6-inch foam thickness tests are shown here. In general, the finer the pore and the thicker the foam, the greater the noise absorption capability.[4]

An important property of industrial component materials today is the level of fire retardency. Many polyurethane foam products are formulated to provide a high level of fire retardency, and such formulations have been successfully tested in accordance with accepted industry standards. It is also important to note that polyurethane foam can be dyed or painted to achieve specific color effects, throughout a wide range, without appreciably reducing its ability to absorb sound energy.

In manufacturing sound-absorbing foams, the variables—including permeability, pore size, thickness and surface finish—are taken into careful

184 INDUSTRIAL NOISE CONTROL HANDBOOK

consideration to produce products that precisely meet specific application requirements. Foams should be selected carefully by industrial designers who have prior knowledge of the frequency range involved, desired thickness, fire retardancy, appearance/color, cleanability, abrasion resistance, and, of course, cost.

USES FOR FOAM IN THE FIELD OF NOISE REDUCTION

If one word were required to describe the essence of this product's contribution to the noise war effort, the word would be "innovation." Flexible polyurethane foam has reduced successfully noise in such current industrial applications as manufacturing plant machinery, metal-removing machinery, industrial vehicle interiors for various uses including agriculture and construction, and large over-the-road trucks. In addition, a growing number of new uses are rapidly gaining acceptance.

Polyurethane foam is being used very successfully in reactive mufflers to further reduce noise from pneumatic control valves. A 1-inch-thick foam disc is placed adjacent to the outlet of the muffler, as part of the unit. This application increases the effective noise reduction of the muffler by an additional 9 dBA.

Another unique and interesting muffler design is found in computer housings, where sound waves pass through access holes into a hollow pillow-like polyurethane foam structure (Figure 9.7). The sound hits two

Figure 9.7 An interesting muffler design is found in computer housings.[5]

internal baffles and then is directed, right and left, out the ends. In all cases, noise levels have been reduced considerably.

Foam also is specified in air-flow acoustical mufflers, especially where space is restricted (Figure 9.8). Air paths can be designed as eccentric curves within foam blocks, for two reasons: (1) to give the noise a longer dissipation path to travel and (2) to cause the noise to impinge on a more absorptive foam area, due to the erratic route. Units currently using this concept have reduced noise levels by 8 to 12 dBA.

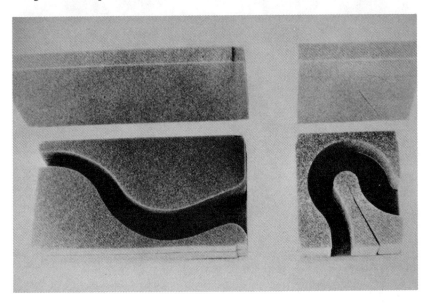

Figure 9.8 Polyurethane foam is used in air-flow acoustical mufflers, especially where space is restricted.[5]

In a large business office, the noise from a copy machine vacuum cleaner disturbed employees working nearby. The installation of a 1-inch-thick ring of very fine pore reticulated polyurethane foam around the outlet of the vacuum pump reduced the noise level by 9 dBA.

Polyurethane foam was recently designed into several components of a complex plastic pelletizing machine. This involved partial modification of the enclosures for the inlet and outlet chutes, the cutting mechanism and the feed mechanism. Noise output from the machine was reduced by 13.5 dBA.

Acoustical wall treatment is a new and rapidly growing architectural application for foam. Materials consist of a fine-pore reticulated foam with a fiberglass woven fabric laminated to it. The treatment is applied like thick wallpaper. At a thickness of $3/8$ inch, the covering has an NRC (Noise Reduction Coefficient) of 0.40, or in effect a 40% decrease in noise. At a 1-inch thickness, it tests at NRC 0.85, or 85% noise reduction. The product in this application was tested in accordance with ASTM-C423 (Figure 9.9).

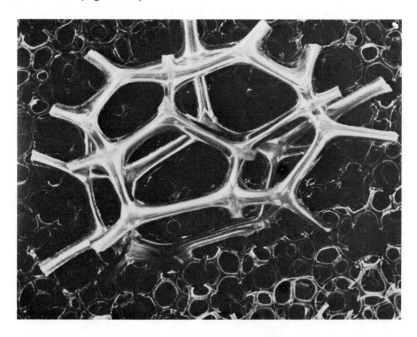

Figure 9.9 One dodecahedral, or 12-faced, cell of Scott's open pore, fully reticulated polyurethane foam, greatly enlarged.

Another similar application is the use of $1/8$-inch foam covered by an open-weave textile material, laminated together in the form of an attractive drape (Figure 9.10). In the completely stretched flat position, noise hitting this drape achieved NRC 0.55. In the completely closed or pleated position, the drape reached NRC 0.65.

Flexible polyurethane foams have found application as components in sound transmission barriers (Figure 9.11). For this use the foam is laminated to a lead-loaded vinyl backing and a perforated textile facing so that it resembles a blanket, in what is called "flexible paneling." The product can be stretched between frames or installed on a wall for modernistic architectural applications, or simply draped, tentlike, over a piece of noisy machinery in an industrial plant. At 250 Hz, using a 0.4 pound-per-square foot vinyl, this blanket application reduced noise by

ADDITIONAL SOUND CONTROL MATERIALS 187

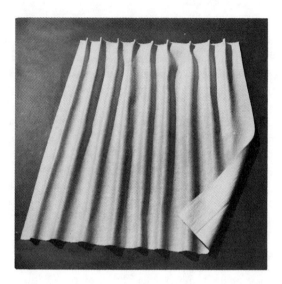

Figure 9.10 Acoustical wall treatment is a new and rapidly growing architectural application for polyurethane foam. One major use involves $1/8$-inch foam covered by an open-weave textile material, laminated together in the form of an attractive drape.[5]

Figure 9.11 Polyurethane foam is being used very successfully in reactive mufflers to reduce noise from pneumatic control valves. A 1-inch-thick foam disc is placed adjacent to the outlet of the muffler, as part of the unit. This application increases the effective noise reduction of the muffler by 9 dBA.[5]

16 dBA. At 250 Hz, using a 0.766 pound-per-square-foot vinyl, it reduced noise 21 dBA. At 1000 Hz, the 0.4 pound-per-square-foot vinyl blanket reduced noise 52 dBA, and the 0.766 pound-per-square-foot vinyl model reduced noise by 59 dBA. Not only is this blanket an effective transmission barrier, but the perforated facing makes it an excellent sound absorber as well.

Open-pore reticulated polyurethane foam has many basic composite forms. Its varied capabilities as a highly efficient sound-absorptive material are well documented, and its versatility in application is limited only by the imagination and creativity of the design engineer. In the responsible analysis of noise problems and their solutions and in virtually any engineering application, open-pore polyurethane foam deserves prime consideration.

AIRCRAFT APPLICATIONS

A polyurethane foam material, specially compressed into sound-absorbing layers, has substantially reduced the noise levels in the two-man crew compartment of the supersonic F-111, the multipurpose military aircraft. The foam serves as an acoustical liner for air inlet ducts to the crew compartment and muffles the noise associated with the plane's air-conditioning system, which supplies the crew with temperature-controlled and pressure-regulated air.

According to engineers of General Dynamics Corp., prime contractor for the F-111, uncontrolled noise levels would affect the efficiency of crewmen and make radio reception difficult. The noise-abating material applied to the ducts is "Scottfelt," which is produced at the Foam Division of Scott Paper Co. by compressing open-pore polyurethane into layers of various thicknesses. The special material was selected by General Dynamics engineers both for its sound attenuation properties and for its resistance to erosion caused by the velocity of air passing through the inlet ducts. These properties were better in Scottfelt than in other materials tested, the engineers said.[6]

The F-111 is the country's first supersonic plane in production that redesigns its wing shape in flight. The variable wing provides stable performance throughout the plane's speed spectrum, from slow approaches to more than twice the speed of sound. The air inlet ducts measure 5 inches in diameter and 30 to 36 inches in length. In addition, there is a 5-foot-long rectangular cross-over duct situated behind the crewmen. This cross-over duct, which is also lined with Scottfeld, receives the incoming air and distributes it in the rear and along the sides of the crew module.

The Scottfelt, in thicknesses of $^1/_4$-, $^1/_2$- and 1-inch, is delivered to the fabricator in sheet form. It is then trimmed to fit the aluminum ducts. For hard-to-reach areas, the foam has a pressure sensitive adhesive backing. In the cross-over duct and other more accessible areas of the air inlet system, the foam is bonded by contact cement.

General Dynamics officials said that noise control is an important part of the environmental control in the F-111. Since the F-111's two jet engines are situated well aft of the crew module, engine noise is not a major problem to the crewmen.

The F-111, in various versions, is capable of performing missions as a tactical fighter, an air superiority fighter, a strategic bomber, a reconnaissance aircraft and a strike aircraft. The plane's variable-sweep wind, fully extended to 16 degrees, creates maximum span and surface area for maximum lift during short takeoffs and landings. As speed increases and lift turns into drag, thus slowing down the aircraft, the span and surface area are reduced by sweeping the wings as much as 72.5 degrees, until the tips rest close to the tail.

USES OF NYLON IN NOISE REDUCTION

Nylon Gears[6]

The noisiest part of any machine is usually gearing simply because of the inherent function of transmitting energy from the power source to the production end of the machine. Metal-to-metal gears clang, whine, and otherwise generate noise that is harmful to a worker's ears. The Polymer Corp., Reading, Pennsylvania, has been producing large nylon gears by the patented MonoCast® process for many years. Originally selected as a long-wearing alternative to bronze, steel and phenolic gearing, the noise-reduction properties of a nylon gear have only recently become an important factor in its selection to replace some other material.[7]

In the past, the paper industry has used more nylon gears than any other industry. They can be found on such machinery as dryers, winders, filters and cutters. The Polymer Corp. is also widely known as a supplier of nylon bearings that replace bronze bearings on overhead cranes.

A large **west** coast mill of a major American paper company was experiencing excessive noise from steel gearing on its high-speed continuous rewinder machines. These high-speed machines, used in the manufacture of bathroom tissue and roll towels, generated gear noise that severely limited the amount of time the operator could stay at his station without the use of ear protectors. In searching for a method of increasing operator productivity without resorting to ear protectors, which are at best a stopgap measure that limits worker effectiveness, the paper mill engineers decided to use nylon for every other gear on the drive train of the rewinding machines. Result: an immediate measurable reduction in the noise level.

To document the tremendous drop in the noise level brought about by the substitution of nylon, paper mill personnel contracted for a noise survey by an acoustical consulting firm in Seattle, Washington. The report describes the measurement technique and presents the results both in dBA and at nine different frequency levels.

Testing Method

Measurements were made using a precision sound level meter and an octave band analyzer. Readings were taken at five different locations around a high-speed continuous rewinder machine with both sound measuring devices. In addition, a tape recorder was used at the operator's position to permit a continuous analysis of the noise spectrum.

At each location the sound pressure level was measured in decibels in octave bands having center frequencies of 31.5, 63, 125, 250, 500, 1000, 2000, 4000 and 8000 cycles per second. In addition, readings were also taken at each location with the sound level meter set at the A-scale. The A-scale frequencies are most important to hearing damage risk and are listed in federal statutes. Measurements were first taken with the rewinder machine running paper and with all-steel gearing. Then, after every other gear on the drive train was changed to nylon, a complete set of readings was again taken at the same locations.

Analysis of Results

In analyzing the effect that the nylon gearing has on reducing noise, Location #1 (the operator's station) and Location #5 (the roll inspector's station) are the most significant. The operator's position is approximately 2 feet from the nearest machine surface and almost any noise reduction here can significantly increase the maximum allowable time the operator can remain at his station without ear protection. In addition to the legal factor, the improvement in working conditions increases worker morale and productivity (see Tables 9.2 and 9.3).

The most significant decrease in decibel level was recorded in the 2000 Hz octave band, which dropped between 9 and 10 decibels while the 1000 Hz band dropped 6 decibels. In this particular instance these are the two frequencies that, when measured on the weighted A scale, have the most effect on raising the dBA.

Table 9.2 Noise Level Readings with All Steel Gears on High-Speed Paper Rewinder[6]

		Octave Band Frequencies							
Location No. 1—Operator's Position									
BAND	dBA	31.5	63	125	250	500	1000	2000	4000	8000
MIN.	100	82	80	84	88	92	95	95	92	83
MAX.	101	89	86	87	91	95	97	96	93	84
Location No. 5—Roll Inspector's Position									
BAND	dBA	31.5	63	125	250	500	1000	2000	4000	8000
MIN.	100	80	80	83	85	91	93	90	85	78
MAX.	98	90	84	87	88	93	95	92	86	85

Table 9.3 Noise Level Readings with Every Other Gear on High-Speed Paper Rewinder Made of Nylon[6]

		Octave Band Frequencies							
Location No. 1—Operator's Position									
BAND	dBA	31.5	63	125	250	500	1000	2000	4000	8000
MIN.	95	91	85	87	86	90	89	85	90	86
MAX.	96	95	89	91	88	92	91	87	92	87
Location No. 5—Roll Inspector's Position									
BAND	dBA	31.5	63	125	250	500	1000	2000	4000	8000
MIN.	90	94	85	86	85	87	85	82	81	79
MAX.	92	99	88	89	87	89	87	84	83	80

With all steel gears the 2000 Hz band had a decibel range of 95-96. When every other steel gear was replaced with a nylon gear, the 2000 Hz band decibel level dropped to 85-87. This reduction eliminated the 2000 Hz band as a prime limiting factor in exposure time. At the same time the decibel level in the 1000 Hz band was reduced from 95-97 to 89-91 decibels, and the 1000 Hz range became the prime limiting factor in maximum allowable exposure.

The most important column on the survey results, however, is the dBA band simply because it takes into consideration the changes in all the frequencies and is the measuring standard set up by the Occupational Safety and Health Act. The dBA at the operator's position dropped from 100-101 with metal gears to 95-96 after every other metal gear had been replaced with nylon (see Figure 9.12). According to law, this raises the time the operator can stay at his position from 90 to 120 minutes with steel gears to 180 to 240 minutes with nylon gears. This 100% increase is very significant when it is considered that it was achieved by changes in only one aspect of the rewinding machine—every other gear on the drive train. No baffles, barriers, or dampening were added and no change in machine speed or output were necessary.[7]

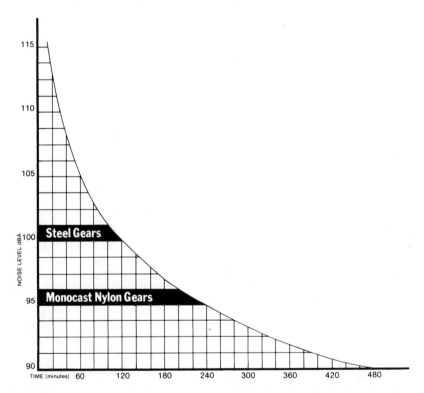

Figure 9.12 The curve represents the maximum allowable exposure time at various noise levels as per the Federal Occupational Health and Safety Act of 1970. With all steel gears on a high-speed continuous paper rewinder, the noise level of 100 to 101 dBA permitted a maximum exposure time of 2 hours. When MonoCast nylon gears were substituted for every other gear on the drive train, permissible exposure time was doubled to 4 hours.[6]

At Location #5, the roll inspector's station, an equally significant change was measured. Here the dBA went from 97-98 with all steel gears to

90-92 when every other gear was changed to nylon. This increased the maximum allowable exposure time at this position from 3 hours to between 6 and 8 hours.

The resiliency of nylon is the material difference that causes nylon gears to produce less noise than steel gears. Sound is a product of vibration, and the metal-to-metal gearing transmits vibration that can become abusive noise. However, because of its physical resiliency, nylon does not produce these same vibrations; hence, there is a quieter machine.

Paper mills have been using nylon gears for many years, and they are rapidly gaining widespread acceptance in other industries. The original reasons for choosing nylon to replace steel, bronze and phenolic—longer wear with less lubrication—have been joined by the newly important property of noise reduction. In the paper industry nylon gears range up to only 72 inches in diameter while a gear that rotates a screening disc for an English power station is 168 inches (14 feet) in diameter. This size is important only to point out the great strides nylon has made as a gear material in recent years.

Keyways

Keyways can be used to drive a nylon gear. With this construction method, however, a check should be made to ensure that the keyway has the required load-carrying capabilities. The keyway always should have a radiused corner to reduce stress concentration. The minimum keyway area is determined from the following formula:

$$A = \frac{20\ HP}{NR}$$

where: A = minimum keyway area (length x height in inches)
HP = horsepower transmitted
N = gear speed in rpm
R = mean keyway radius in inches

(See Figure 9.13.)

Set Screws

Set screws can also be utilized in nylon gears, but consideration must be given to the construction and the type of service required. In the event that severe service may be encountered, it is suggested that a metal hub be used.

On installation, care should be taken when tightening the set screw. The maintenance mechanic will realize a different "feel," and may, therefore, overtighten the screw. It is suggested that the set screw be hand tightened, then tightened one-half a rotation.

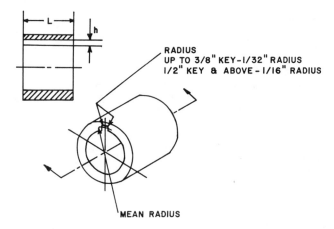

Figure 9.13   Keyway used to drive a nylon gear.

Metal Hubs

For Nylatron gearing that is over 8 inches in diameter, or for gears that will see relatively severe service, it is suggested that a metal hub be utilized.[8]

PVC FILM NOISE CONTROL[3]

Polyvinylchloride is another material used by industry for noise attenuation. In the form of a transparent noise barrier material that is limp and flexible, PVC film resists the passage of sound waves and reduces noise transmission. It permits a machine operator, for instance, to monitor his machine even though it may be surrounded by an acoustical barrier, such as a curtain with one or all panels made of this material. It reduces noise by an average of 20 dB (STC). Made of tough, tear-resistant cast PVC, it is resistant to yellowing, fading, and clouding in normal industrial applications. PVC film is available in 20-yard rolls, 54 inches wide, 0.070 inches thick, and weighing $1/2$ lb/sq ft. This material is recommended for use as hanging curtains and as "windows" in curtains and fabricated enclosures for noise control systems where visual monitoring is desired.

Typical enclosure-type applications include: presses, machine tools, textile looms, twist frames, conveyor lines, packaging lines, office equipment, control panels. It can be used alone or as window panels in opaque noise control systems. Transmission loss test data can be seen in Figure 9.4 and the film itself is illustrated in Figure 9.15.

ADDITIONAL SOUND CONTROL MATERIALS 195

NOISE CONTROL PERFORMANCE

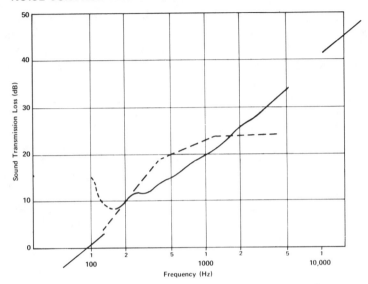

Figure 9.14   Transmission loss test data for PVC film.[3]

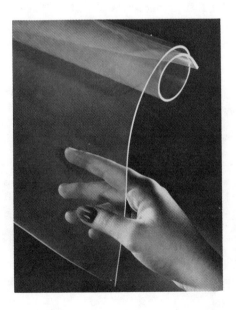

Figure 9.15   PVC film used in noise control.

FIBER-LOADED VINYL[3]

Fiber-loaded vinyl is a flexible noise barrier material that is a limp, tough, high-temperature-fused vinyl, loaded with high-density, nonlead fillers and reinforced with high-strength fabric. It resists the passage of sound waves and reduces noise transmission because it doesn't resonate the way metal mass barriers do.

Available in standard $1/2$- and 1 lb/sq ft densities with acoustical ratings of STC 20 and 27, respectively, it can be used in layers to achieve higher STC levels and follow the limp-mass loss principles of noise transmission. This material has high tear and tensile strengths, it will not rot, shrink, or cause metal corrosion, and it is easy to fabricate and install because it can be cut with a knife or scissors, or die-cut. It is also safe and nontoxic.[3]

It is available in a variety of colors, with textured or embossed surfaces, with Saran coating for USDA-approved installations in food processing plants. This material has been confirmed by plant engineers, architects, and design engineers to be an effective noise barrier in industrial, OEM, new-construction, and remedial applications such as rollaway curtains, pipe- and duct-noise lagging, machinery covers, ceiling noise barriers, crosstalk barriers, wall and door septums, noise panels, and rooftop-equipment barriers. Samples of this product are shown in Figure 9.16.[3]

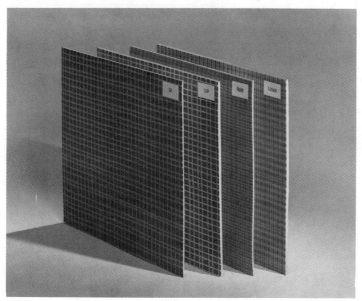

Figure 9.16 Fiber-loaded vinyl.

LEAD-LOADED VINYL

There are many lead-loaded vinyls employed by industry for sound control, one of which is Ferro Coustifab produced by Ferro Corp., Norwalk, Connecticut, illustrated in Figure 9.17. This lead-loaded vinyl fabric is a

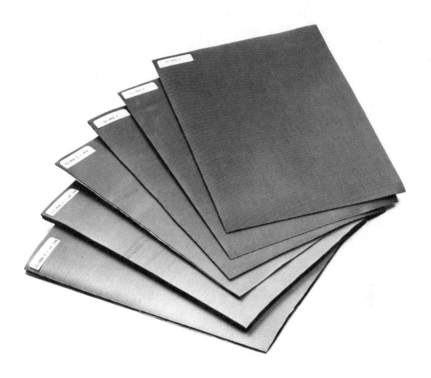

Figure 9.17 Lead-loaded vinyl.

fiberglass-reinforced material coated with lead-loaded vinyl for the reduction of noise transmission, ranging from 20 to 29 dB (STC). Higher STC levels can be achieved with multilayers of this material. Made with vinyl for toughness, limpness, durability, appearance, and oil and chemical resistance, it is reinforced with fiberglass for strength, durability, and flame resistance. It is loaded with lead powder for weight (limp mass) and maximum noise attenuation with minimum thickness.

Lead-loaded vinyl is available in over 15 industrial and aircraft grades, with densities ranging from 0.14 to 1.50 lb/sq ft. It is also used with polyurethane foam for improved sound transmission loss, and with aluminum foil to provide reflective thermal insulation and a noise barrier in one material for pipes and ducts.

It is easy to fabricate and install because it can be cut with a knife or scissors, or die-cut. Approved by FDA for use in food and drug plants, the fabrics reduce airborne and structure-borne noise. It can be used alone, in layers, or in combination with acoustical glass, wool, or sound absorption foam. Typical applications for the use of this product include: aircraft sound blankets, gearbox pillows, air-intake-duct bellows, insulated pipeline lagging, door and panel barrier diaphragms, noise isolation enclosures, impact or sympathetic vibration noise dampers. It is not recommended for dynamic movable curtain applications requiring USDA approval or air handling ducts and plenums where building codes require UL 181 air-handling specifications.

DAMPING COMPOUND[7]

There is also a compound used for reducing excessive noise and vibration that meets noise control levels under the Occupational Safety and Health Act (OSHA 1910.95 and others). It is called Flexane, and is highly effective for the control of noise caused by the impact of one material upon another (for example, steel on steel) and for damping sound generated by vibrating metals and nonmetallic plates or panels. It is easy and fast to use because it can be applied by brush or putty knife directly onto existing equipment. In some cases this work can be done without shutting down production. Figure 9.18 illustrates an application of this material.

Another application is the lining of feed hoppers used by large automotive companies. Castings vary in weight from 1 to 10 lbs each, with approximately 2000 lbs of castings dumped in a series of hoppers three to five times each hour, 24 hours a day, 6 or 7 days per week. Flexane was used by the same company for lining metal chutes, gondola tippers or dump boxes, Shaker conveyors and Syntron bowl feeders. The primary reasons for using Flexane are its high abrasion resistance, the reduction of noise level to acceptable limits, its easy applicability, and its lower cost than other materials or methods (see Figure 9.19).[7]

Damping Compound Case History

Damping compounds in paste and sheet form are in wide use for reducing noise and vibration on sheet metal and similar thin materials. One such damping compound is the DYAD polymer by Soundcoat Co., Inc., New York, New York. This compound provides a solution to noise

ADDITIONAL SOUND CONTROL MATERIALS 199

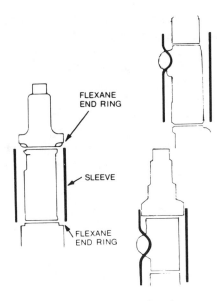

Figure 9.18 Flexane was used for casting rings around cylinders of pneumatic hammers used in underground zinc mines.[7]

Figure 9.19 Lining vibrating bowl feeder with Flexane 85, which is also used on vibrating chutes and screens.[7]

problems associated with subway car wheels, railroad bridges, transformer housings, transmission gears, metal conveyors, punch presses and even high-speed saw blades.

One particular application that provides evidence of DYAD's ability to reduce noise levels is that of the subway car wheel (Figure 9.20). Soundcoat Company research indicated that the high screech associated with a turning subway car is caused by vibration of the wheels. This vibration is excited by the friction of wheel against track, but it is the wheel that makes the noise, not the track. After careful tests and evaluations, it was determined that the DYAD treatment on the 5-inch-thick steel forged wheel provides over 35 dB reduction of wheel screech when the wheel is excited in sharp turns. While each wheel weighs 550 pounds, the treatment weighs a total of 23 pounds—the DYAD and the steel constraining layer.

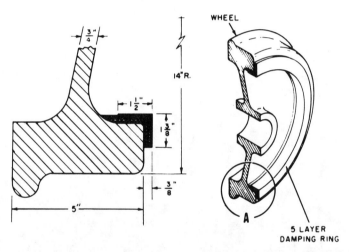

Figure 9.20 Application of a damping compound on a subway car wheel.[2]

Figure 9.21 graphically illustrates the effectiveness of DYAD in reducing noise on a subway car wheel going around a 90° curve. To achieve these results, the constraining layer should be at least one-tenth the thickness of the base structure. For better damping response across a broader frequency range, it is suggested that a constraining layer of one-fourth to one-half the base structure thickness be used.[2]

Epoxy and contact adhesives can be used to join the DYAD to the structure and the constraining layer, provided clean surfaces are prepared and sufficient pressure is applied during curing. Mechanical or welded fastenings are also satisfactory.

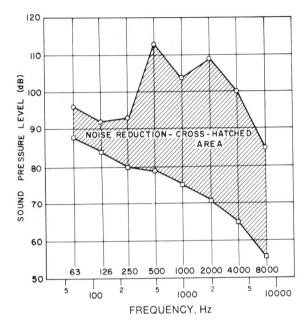

Figure 9.21 Effectiveness of a damping compound on a subway car wheel.[2]

REFERENCES

1. "Foam Reliability in Noise Control," Delaware Technical Report #224, Specialty Composits Corp., Delaware Industrial Park, Newark, Delaware.
2. Soundcoat Company, Inc., Brooklyn, New York 11201.
3. Ferro Corporation, Composites Division, Norwalk, Connecticut 06852.
4. Pizzirusso, J. Chief Technical Consultant, Scott Paper Company, Foam Division, Chester, Pennsylvania 19013.
5. Scott Paper Company, Foam Division, Chester, Pennsylvania 19013.
6. Polymer Corporation, Reading, Pennsylvania 19603.
7. Devcon Corporation, Danvers, Massachusetts 01923.

CHAPTER 10

SILENCERS AND SUPPRESSOR SYSTEMS

Of the variety of silencers industry employs for noise attenuation, one is a duct silencer, which is designed to accommodate air flow in either direction of the duct. Figure 10.1 illustrates a typical duct silencer, the top and bottom of which are filled with a sound-absorbing inorganic material and covered with perforated galvanized steel sheeting. Figure 10.2 illustrates a circular duct silencer, which is filled on the inside with an inorganic material along the exterior walls and the inner shaft; this material is covered with galvanized steel.

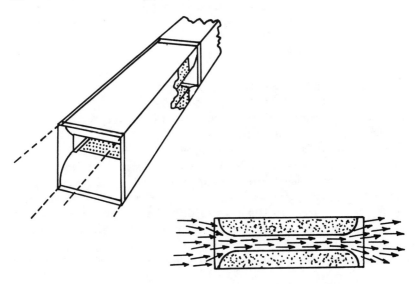

Figure 10.1 Typical duct silencer.

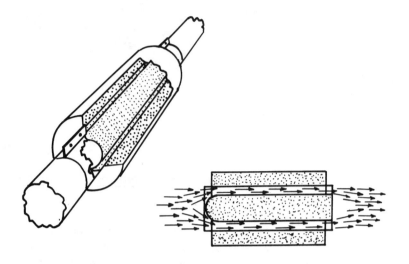

Figure 10.2   Circular duct silencer.

JET ENGINE TESTING[1]

Silencers are used in many applications. Figure 10.3 shows vent air intake silencer filter units that are installed at engine test cells. Similar silencer systems were used at Airesearch Manufacturing Division, Los Angeles, California, where the systems were installed in an aircraft turbo fan engine test cell. After the silencer system was installed, microphones were positioned in the test cell, as seen in Figure 10.4, and also around the exhaust silencer and the inlet silencer. These microphones were hooked up to a tape recorder and the output performance points were recorded; instrumentation data is illustrated in Table 10.1.

The amount of noise reduction achieved by the exhaust and inlet silencers was determined by calculating the difference in dB level between the microphones within the test cell and the average of the microphones outside the cell near the silencers. The range of attenuation values for the engine operation from 1000 pounds to 4100 pounds thrust are illustrated in Figure 10.5; for the inlet silencer in Figure 10.6; for the exhaust silencer systems in Figure 10.7.

The arithmetic average of the values at each octave band of Figures 10.5 and 10.6 were determined in order to obtain single attenuation values for the inlet and exhaust. Figure 10.8 shows the average inlet noise reduction as compared with the required acceptable levels; Figure 10.9 illustrates

Figure 10.3 Vent air intake silencer. Filter units are installed at McCulloch Corporation engine test cells.[1]

206 INDUSTRIAL NOISE CONTROL HANDBOOK

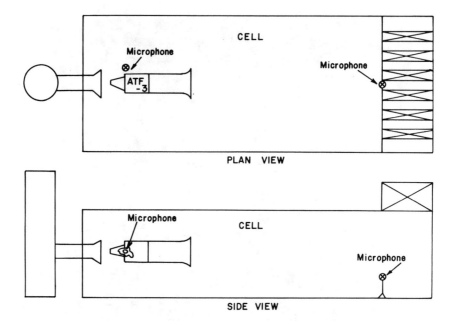

Figure 10.4   Positioning of microphones in the test cell.

Table 10.1   List of Data Acquisition and Reduction Instrumentation

Data Acquisition

Microphone	B & K Type	Preamplifier
1	4134	2615
2	4134	2615
3	4134	2615
4	4134	2615
5	4135	2615
6	4135	2615

Calibrator: B & K Type 4220 Pistonphone
Power supplies: B & K Type 2803
Tape recorder: Honeywell Model 7600

Data Reduction

Tape recorder: Honeywell Model 7600
Octave filter set: B & K Type 1612
Graphic level recorder: B & K Type 2305

SILENCERS AND SUPPRESSOR SYSTEMS 207

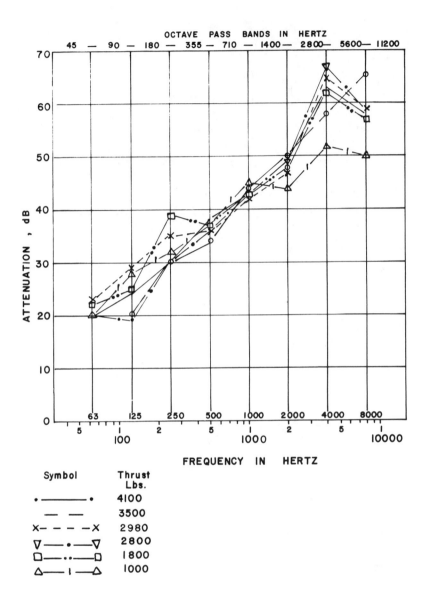

Figure 10.5 Range of attenuation values for the inlet silencer.

208 INDUSTRIAL NOISE CONTROL HANDBOOK

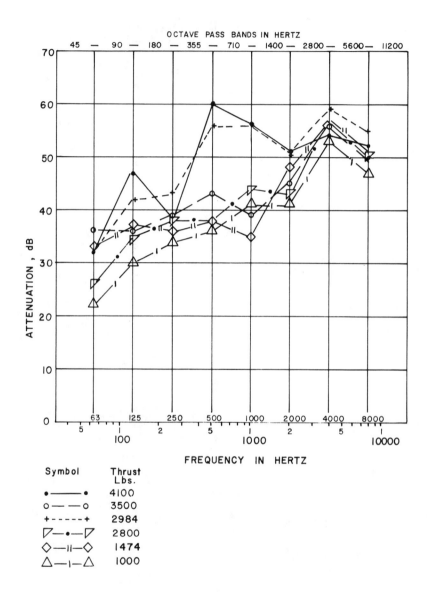

Figure 10.6 Range of attenuation values for the outlet silencer.

Figure 10.7 GAC Model 1260 intake and exhaust silencer systems for ATF-3 through ATF-6 fanjet engine production testing.[1]

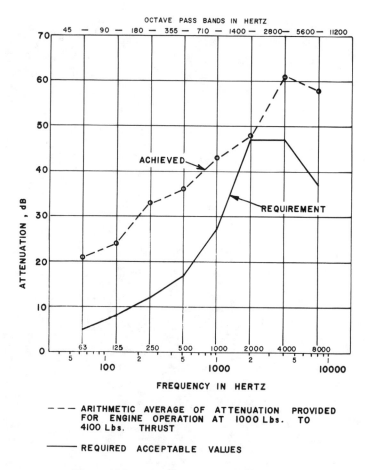

Figure 10.8 Average inlet noise reduction.

the same data for the exhaust silencer. Upon inspection of these two figures, one can see that the inlet and exhaust silencers performed satisfactorily. To illustrate the total noise attenuation acceptability of the inlet and exhaust silencers, Figure 10.10 shows the noise coming from the system compared with acceptable levels in residential surroundings. Figure 10.11 shows an overall view of the total noise suppressor system.

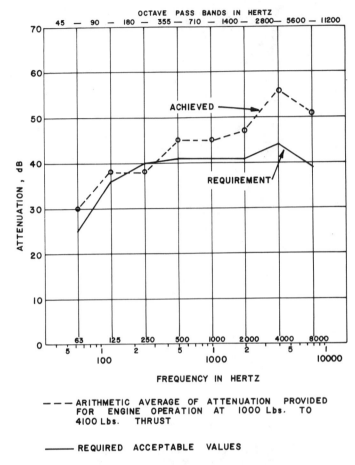

Figure 10.9 Average exhaust noise reduction.

AIRCRAFT NOISE SUPPRESSOR SYSTEMS

Demountable noise suppressors and demountable test cell noise suppressors are used widely by the U.S. Air Force for noise control. A proven noise suppressor system designed by General Acoustics Corporation, Los Angeles, for the Northrop F-5E International Fighter and F-5A & B aircraft and the General Electric J85-13 & -21 engines will be discussed here. Figure 10.12 illustrates a demountable noise suppressor system that has met U.S. Air Force specifications. Use of this system has led to greater hearing protection for technical personnel, who now show a higher working

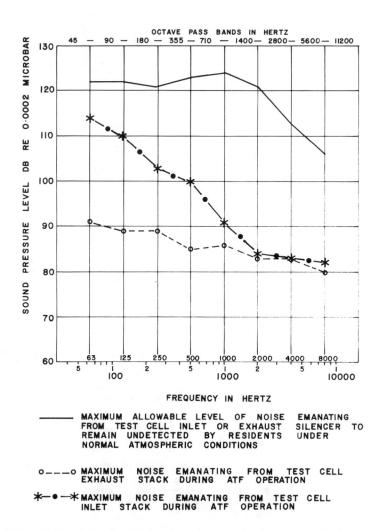

Figure 10.10 Noise from the entire system compared to acceptable levels.

efficiency and morale. Figure 10.13 shows a side view of the suppressor system that has proven itself most efficient in every environment.

Test cells, like the one illustrated in Figure 10.14, are made from prefabricated modular structures and are used by the Air Force. These cells are used for testing engines such as the ATF-3, -4, -5, and -6; JT-3D, and JT-8D. It was found in the test cell installation that a conventional water spray could not be used for cooling because the water spray mixed

SILENCERS AND SUPPRESSOR SYSTEMS 213

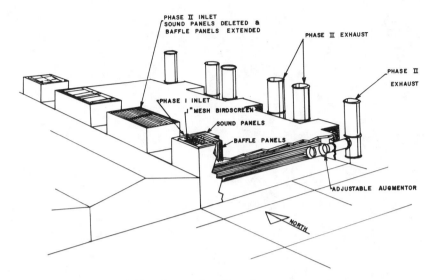

Figure 10.11 Surrounding view of suppressor system.

Figure 10.12 Demountable noise suppressor system.

214 INDUSTRIAL NOISE CONTROL HANDBOOK

Figure 10.13 Side view of demountable suppressor system.

Figure 10.14 Test cells, like the aircraft noise suppressors, are made from prefabricated modular structures. Such construction ensures ease of installation at less cost in time and money and provides for expansion or relocation if required. Similar test cell noise suppressor systems have been made by General Acoustics for the ATF-3, -4, -5 and -6, JT-3D and JT-8D engines as well.

with the exhaust gases, yielding an acid mixture that would corrode the aircraft surfaces. However, as an alternative, a closed loop recirculating water cooling system was developed and installed in this cell. Figure 10.15 illustrates a complete suppressor system; Figure 10.16 shows the positioning of a jet during a test.

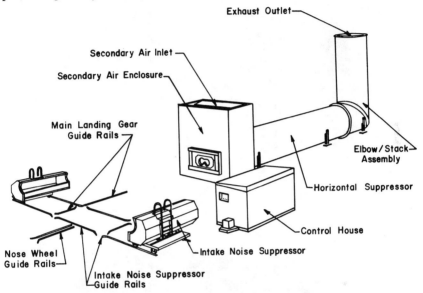

Figure 10.15 Breakdown of complete suppressor system.

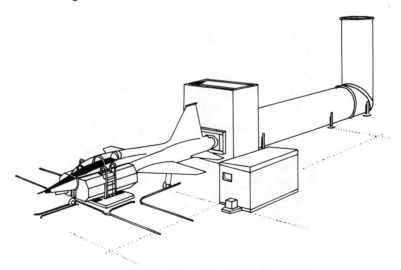

Figure 10.16 Jet positioning during a test.

TYPES OF INDUSTRIAL SILENCERS[2]

Figures 10.17 and 10.18 illustrate two types of silencers used in industry for many kinds of machines. Compact silencers are shown in Figures 10.19-10.22. Figure 10.19 shows how the packaged equipment must be

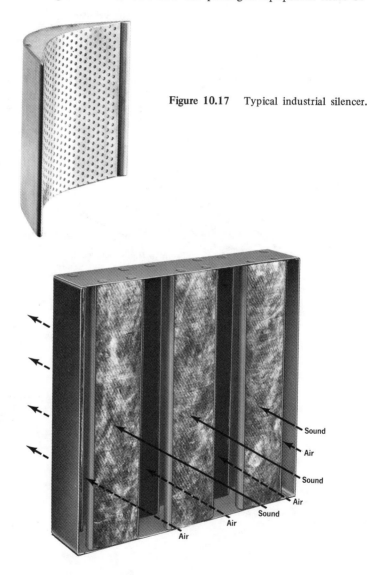

Figure 10.17   Typical industrial silencer.

Figure 10.18   This silencer illustrates how air and sound enter its construction and only air leaves.

SILENCERS AND SUPPRESSOR SYSTEMS   217

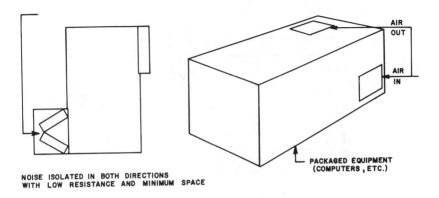

NOISE ISOLATED IN BOTH DIRECTIONS
WITH LOW RESISTANCE AND MINIMUM SPACE

Figure 10.19   Lining packaged equipment stops vibration noise.

lined to eliminate transmission of noise through vibration. In Figure 10.20 we see a compact silencer used between transfer grilles to cut down cross talk and room-to-room noise transfer. Figure 10.21 shows a compact silencer used in return air ducts behind a return grille to reduce equipment noise. Figure 10.22 illustrates a compact silencer applied in a suspended acoustical ceiling, with troffers supplying air to a given area. The area above the suspended ceiling is used for return air. This space has long been a problem because of the cross transfer of noise from one area even though the areas have been partitioned off for privacy.

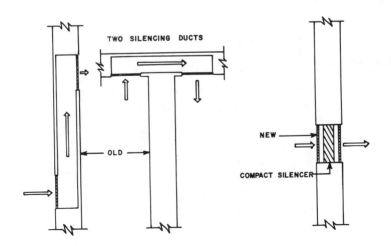

Figure 10.20   Silencer between transfer grilles.

218 INDUSTRIAL NOISE CONTROL HANDBOOK

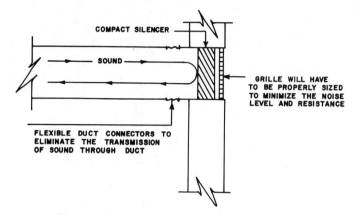

Figure 10.21   Silencer used behind return grille.

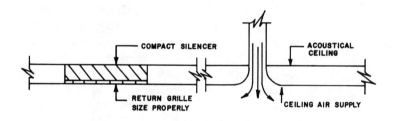

Figure 10.22   Silencer used in a suspended ceiling.

REFERENCES

1. General Acoustics Corporation, Los Angeles, California.
2. Air Filter Corporation, Milwaukee, Wisconsin.

CHAPTER 11

FUNDAMENTALS OF VIBRATION*

NATURE OF VIBRATION

Vibration may be considered as an oscillating motion of a particle or body about a reference position. This motion can be periodic, random or transient. Simple examples of each will be discussed here.

Periodic Functions

The sinusoidal signal of Figure 11.1 is by definition at one discrete frequency, which is given by 1/T, and has the units of Hz (Hertz), which is the number of cycles/second. T is the time taken for the wave to perform one complete cycle. This motion is periodic in that it repeats itself at regular intervals of time (T). The nearest that one would come to a pure tone in vibration is the movement of the arms of a tuning fork.

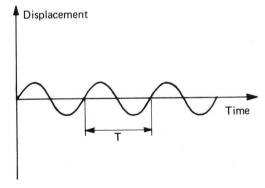

Figure 11.1 Example of a simple harmonic (sinusoidal) vibration signal (f = 1/T).

---

*By Anthony J. Schneider, B & K Instruments, Inc., Cleveland, Ohio.

The vibration of Figure 11.2 is still periodic because it repeats itself at regular intervals, but it is now complex (*i.e.*, it is not purely sinusoidal). In fact, this particular example is a combination of two sine waves and is the sort of vibration produced during the piston acceleration of an internal combustion engine. A periodic signal need not be symmetrical, and it can be made up of many combinations of frequencies of different amplitudes and phase.

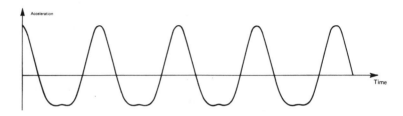

Figure 11.2 Example of compound harmonic periodic motion (f + 2f).

The signal of Figure 11.2 is made up of two components of different frequencies and amplitudes, and these are shown in Figure 11.3. In this example one sine wave has a frequency that is twice the other, and its amplitude is much smaller. The amplitudes are summed arithmetically at each point in time to obtain the combined signal level.

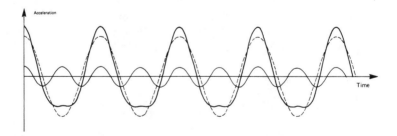

Figure 11.3 Waveform of Figure 11.2 split into its components.

An important and useful way of portraying the amplitudes and frequencies of all the different components of a complex signal is the frequency spectrum, which is a plot of amplitude against component frequency. A few simple examples are given in Figure 11.4.

As can be seen from Figure 11.4A, the simple sine wave has a spectrum of one discrete frequency (f = 1/T), and is represented by a line whose

FUNDAMENTALS OF VIBRATION 221

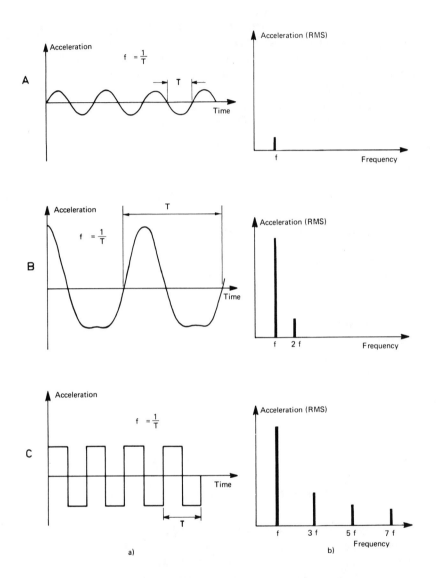

Figure 11.4 Examples of (a) periodic signals and (b) their frequency spectra.

height is proportional to the amplitude of the wave. The waveform of Figures 11.2 and 11.3 has two components, each having a discrete frequency. One frequency is twice the other. The two frequencies each have one line on the spectrum, whose height is proportional to amplitude

(Figure 11.4B). In this case, the frequency f would be called the fundamental frequency and the frequency 2f is its second harmonic, since it is twice the frequency of the fundamental. A "square" wave, shown in Figure 11.4C, is another example of a periodic function. This consists of a fundamental frequency f, and *odd* harmonics only (*i.e.*, 3, 5, 7f) that drop off in level at a defined rate.

All periodic functions, such as those shown above can be defined by precise mathematical equations, which can make their analysis much simpler.

Random Vibration

The most commonly encountered type of vibration in everyday life is random vibration. It is continuous, but nonperiodic, and contains many frequency components. While many of these components will be related in the form of harmonics of certain frequencies due to machinery movements, many components will be entirely independent of these.

A random vibration signal is shown in Figure 11.5. It lacks repetition, its frequency spectrum is broad band or continuous, since it contains all frequencies, and it has some spikes due to resonance or to harmonics of certain vibrating components. These spikes are often caused by the main source of vibration, and once they are discovered from frequency analysis they are relatively easy to isolate.

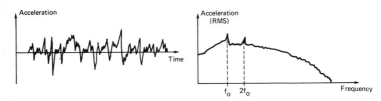

Figure 11.5 Example of random vibration and a typical frequency spectrum.

A random vibration is by definition nondeterministic; it can be represented mathematically only by a series of probability statements, since to obtain a complete description of the vibrations an infinitely long time record is theoretically necessary. A random vibration is called "stationary" if successive samples taken of it are essentially the same in character, although the rigorous mathematical definition of this is somewhat more complicated. A stationary random vibration is normally easier to analyze than a nonstationary one.

Transient Vibration

A transient vibration is noncontinuous; it occurs due to impact, or during the starting up of a motor, or anywhere that the exciting force is not continuous. A transient vibration is often deterministic.

The transient vibration shown in Figure 11.6 has a spectrum with many "lobes," as shown, that would contain a whole range of frequencies. The size of the lobes (in terms of frequency range) will depend on the duration of the transient, and their relative amplitudes will depend on the shape of the transient itself.

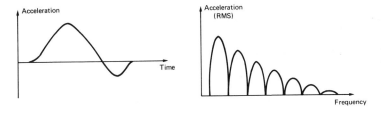

Figure 11.6   Example of a transient vibration and a typical frequency spectrum.

SOURCES OF VIBRATION

Vibration results from some part of moving machinery being out of balance, turbulent fluid flow, rattling of loose objects, impulses, shocks—the list is endless. Vibration is normally undesirable, but it can often be reduced by careful design or by development modifications. Sometimes, however, vibration is introduced to make things work, such as vibrating conveyor belts, mechanical hammers and even musical instruments. In such cases resulting noise is unavoidable. It is then important to isolate it as much as possible, even to the extent of preventing your hi-fi loudspeakers from rattling your neighbor's picture on your adjoining apartment wall!

EFFECTS OF VIBRATION

The effects of vibration are often serious. Humans subjected to vibration can be affected by blurred vision, loss of balance and consequent lack of ability to do their job properly. In some cases, certain frequencies and levels of vibration can permanently damage internal body organs. Machinery can also be damaged by vibration. If the vibration occurs at the resonance frequency of some component it can be cracked or broken by fatigue, and nuts, bolts and rivets can be shaken apart. Noise resulting from vibration is also often a serious problem, and can be a health hazard to people exposed to it for long periods.

One really difficult thing about vibration is that it will not stay in one place unless special steps are taken to isolate it. Vibration is transmitted through any solid object in contact with it, including the floor, walls, pipes, electrical conduits and any other mechanical linkages, which in turn cause things in contact with them to vibrate and radiate noise. Even if a particular frequency from one machine does not cause any of its own components to resonate, it could quite possibly excite a resonance in a connected machine. It is thus important to measure and control vibration.

MEASUREMENT OF VIBRATION

It is necessary to measure vibration for several reasons. While sometimes the vibration picture is obvious, there are many occasions when it is not, and it would be senseless to design a means of reducing a harmonic component frequency while vibration continues at the fundamental frequency. From the design point of view, the magnitude and frequency of the vibration need to be known to ensure that the stresses induced are not too great for the material to withstand. Another important reason is that if a certain piece of machinery has been found to resonate at a given frequency, the operator can avoid running the machine at a speed that will excite that resonance. For vibration damping or isolating, it is necessary to know the amplitudes and frequencies involved in order to select the correct damping materials. Preventive maintenance is another area where vibration monitoring is useful, since many faults (like worn teeth on a gear wheel or the start of a roller bearing's failure) can be detected long before failure merely by noticing changes in the vibration spectra.

There are three main parameters used to describe vibration: displacement, velocity and acceleration. Displacement is the distance moved by the measuring point from its natural position, velocity is the speed at which that point moves, and acceleration is the rate of change of its speed with time. Displacement is proportional to strain in the material. Velocity is a function of kinetic energy that must be dissipated in the machine. Therefore, velocity is related to potential damage in the machine.

Acceleration is proportional to the force acting on an object. These parameters are measured in the normal units of distance, velocity and acceleration, which are meters (m), meters per second (m/s) and meters per second squared (m/s^2), respectively. The units are in accordance with ISOR 1000 (SI units). Acceleration is also often measured in terms of gravitational constant (g), and it is common to see vibration ranges or levels quoted in terms of numbers of g because it is easy to calibrate measuring equipment at levels of 1 g, and also because g is an internationally understood symbol, regardless of units.

Historically, displacement was the first parameter to be measured because with slow-moving machines and large displacements it was the easiest to see and was measurable by simple optical methods. Displacement limits were then set by the law of elasticity, the fatigue of materials or by mechanical clearances. Measurement of displacement emphasizes very low frequencies. At higher frequencies of vibration, displacement is negligible.

With the introduction of higher-speed machines, the displacements became smaller and more difficult to see, but breakdown could still occur, so a move towards velocity measurements was made. Many rotating machines have a vibration frequency spectrum with a fairly even distribution of energy across the frequency range up to 1 kHz (Figure 11.7), thus making it reasonably easy to prescribe a velocity limit for a particular type of machine. (It is even easier to prescribe for single resonance

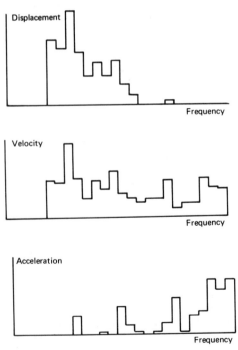

Figure 11.7 Spectra showing the effects of measuring displacement, velocity and acceleration on rotating machinery in the frequency range up to 1 kHz.

frequencies). Standards organizations in several countries now recommend this type of limit especially for installed electric motors. Measurement of acceleration puts emphasis on the higher frequencies, where it is often important to understand the vibratory forces acting on the machine.

For high-frequency vibration (above about 1 kHz), it is necessary to measure acceleration, because displacement velocity drops off rapidly at high frequencies. The reason for this will be seen later. Since high frequencies play a large part in preventive maintenance, early detection of breakdowns and vibration monitoring acceleration measurements are very important. For example, many faults in machine elements, like roller bearings, can be detected in the frequency range 20 kHz-50 kHz long before failure occurs, thus making it possible to plan repairs and shutdown times.

RELATIONSHIP BETWEEN DISPLACEMENT, VELOCITY AND ACCELERATION

There is a well-defined relationship between displacement, velocity and acceleration. The concept of simple harmonic motion serves as a basis for discussing the mathematical relationships among these parameters.

Figure 11.8 shows a wheel rotating at a constant angular velocity of ω radian/s; θ is measured counterclockwise from the bottom, and x is measured to the right. (There are 2π radians in one revolution, so the wheel is turning at $\omega/2\pi$ rev/s, or $\omega/120\pi$ rev/min.) The wheel drives a

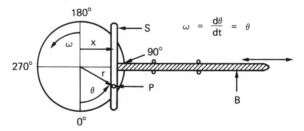

Figure 11.8 Example of simple harmonic motion.

reciprocating mechanism B by means of a protruding stud P, which is free to move in the slider S. B is said to move backwards and forwards with simple harmonic motion (provided there is no mass or friction in any of the sliding contacts) in the horizontal plane. Thus the displacement (x) of B will be as shown on the curve, and one cycle will be completed with each complete revolution of the wheel. In this case revolutions/second ≡ cycles/second ≡ Hz, and the standard relationship between angular speed ω

and frequency f is obtained:

$$\omega = 2\pi f$$

The displacement x of B from its central position is given by:

$$x = r \sin \omega t$$

when $\theta = 0°$ or $180°$, $x = 0$; $\theta = 90°$, $x = +r$; $\theta = 270°$, $x = -r$.
The maximum displacement is given when $\sin \omega t = \pm 1$, and is thus equal to r.

Differentiating displacement with respect to time, we obtain the velocity v of B:

$$v = \frac{dx}{dt} = \dot{x} = r\omega \cos \omega t$$

when $\theta = 90°$ or $270°$, $\dot{x} = 0$; $\theta = 0°$, $\dot{x} = +r\omega$; $\theta = 180°$, $\dot{x} = -r\omega$.
The maximum velocity, V, occurs when $\cos \omega t = \pm 1$, and $V = r\omega$.

Differentiating again with respect to time, we obtain the acceleration a of B:

$$a = \frac{dv}{dt} = \frac{d^2x}{dt^2} = \ddot{x} = -r\omega^2 \sin \omega t$$

when $\theta = 0°$ or $180°$, $\ddot{x} = 0$; $\theta = 90°$, $\ddot{x} = -r\omega^2$; $\theta = 270°$, $\ddot{x} = +r\omega^2$.
The maximum acceleration, A, occurs when $\sin \omega t = \pm 1$, and $A = r\omega^2$.
Figure 11.9 illustrates the effects of these relationships.

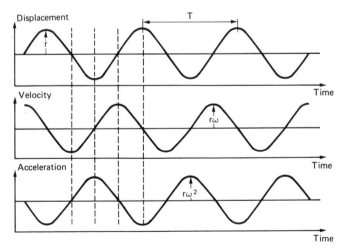

Figure 11.9 Phase and amplitude relationships between displacement, velocity and acceleration.

It can be seen from Figure 11.9 that there is a phase difference between these components. The velocity leads the displacement by 90° (or one quarter of a cycle), the acceleration leads the velocity by 90°, and so the acceleration leads the displacement by 180°. However, in terms of measurement this phase difference does not matter because RMS or peak reading are usually obtained on a meter or recorder.

It can also be seen that each differentiation has multiplied the signal by ω, which does have practical significance. It means that at higher frequencies, the acceleration signal is the largest by a factor of ω or ω^2. Since $\omega = 2\pi f$, the acceleration signal is often the easiest one to measure, or sometimes the only one that can be measured, because the other signals drop into instrumentation noise.

The main significance of the relationship between these paramters is that at any given frequency one measurement can provide the other two. Sensors having a flat acceleration response (accelerometers) are the most common vibration sensors. Integrating the acceleration will give velocity, and integrating again will give displacement, and this is a fairly standard procedure. This integration is performed by use of a low pass filter with a roll-off of $1/f$ for velocity and $1/f^2$ for acceleration.

The emphasis of different parts of the spectrum can be seen quite clearly from the spectra of Figure 11.7. In this case we have a fairly "flat" velocity spectrum. The division by ω to obtain displacement attenuates high frequency signals (at high frequencies ω is large), and in this case the signal disappears. The multiplication by ω to obtain acceleration emphasizes high-frequency signals. Thus, while acceleration may not be the parameter we wish to study at high frequencies, it is the one we would measure in order to obtain velocity.

When drawing spectra, we normally plot the frequency on a logarithmic scale, as it has the effect of expanding the lower frequencies and compressing the higher frequencies, and allowing us to have a reasonable resolution without large sheets of graph paper. Logarithmic plotting also ties in well with the concepts of an octave (doubling of frequency) and a decade (multiplying the frequency by 10) because one octave is now a constant width on the paper, as is one decade, regardless of the frequency range it is covering. Linear scales are useful sometimes, particularly for sorting out harmonics.

We also commonly use a logarithmic scale to plot vibration amplitudes. This gives us the decibel (dB) scale, which is also commonly used in acoustics, electronics and control theory. Part of the use of the two logarithmic scales of level and frequency can be seen in Figure 11.10 in which all the parameters now have straight line relationships with each other for one constant parameter. In this case (Figure 11.10) we have a

FUNDAMENTALS OF VIBRATION 229

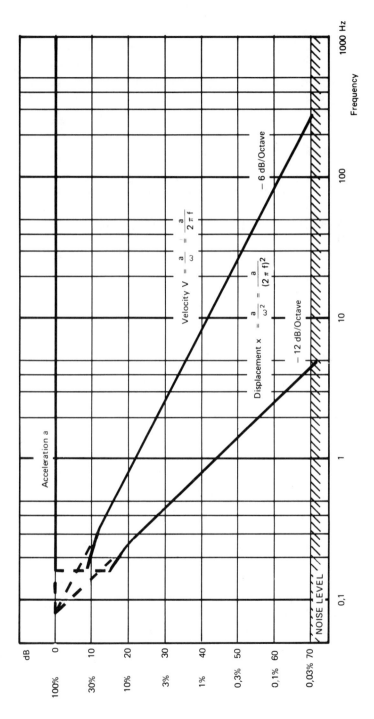

Figure 11.10 Relationship of levels of displacement, velocity and acceleration.

flat acceleration spectrum, and the velocity and displacement drop off with frequency at rates of 6 dB per octave (20 dB per decade) and 12 dB per octave (40 dB per decade), respectively. The displacement drops into instrumentation noise very rapidly, and the velocity signal a little later. At very low frequencies, distortion and cut-off is introduced by the limitations of the measuring and integrating equipment.

PEAK, AVERAGE AND RMS VALUES

There are several different ways of quantifying the level of vibration. The first is to use the peak (or maximum) value, as was done with the analogy of a turning wheel earlier (Figure 11.8). This peak value is shown in Figure 11.11. It is useful for simple harmonic vibration (such as the one shown), but for other types it is not as useful because it depends only on an instantaneous vibration magnitude, and takes no account of time history producing it.

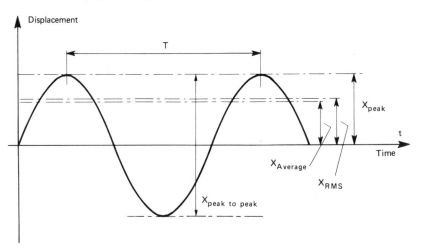

Figure 11.11 Peak, average and RMS values for a sine wave.

The peak-to-peak value, or the magnitude of the positive and negative extremes of the motion, is also commonly used. For a symmetrical signal it is twice the peak value. Another quantity, which does take into account the time history, is the average absolute value, which is defined as

$$x_{|average|} = \frac{1}{T} \int_0^T |x| \, dt$$

Even though this quantity takes into account the time history over a period T, it has been found to be of limited practical interest because it has no direct relationship to any useful physical quantity. A much more useful descriptive quantity, also taking the time history into account, is the RMS (root mean square) value, defined as

$$x_{RMS} = \sqrt{\frac{1}{T} \int_0^T x^2(t)\, dt}$$

The major reason for the importance of the RMS value is its direct relationship to the energy content of the vibrations.

The relationship between these values can be expressed as

$$x_{RMS} = F_f x_{|average|} = \frac{1}{F_c} x_{peak}$$

where F_f and F_c are called form factor and crest factor, respectively, and give some idea of the waveshape of the vibrations being studied.

For a sine wave, this will be

$$x_{RMS} = \frac{\pi}{2\sqrt{2}} x_{|average|} = \frac{1}{\sqrt{2}} x_{peak}$$

motion, $F_f = \dfrac{\pi}{2\sqrt{2}} = 1.11$ ($\cong 1$ dB) and $F_c = \sqrt{2} = 1.414$ (= 3 dB)

Most vibrations encountered will not be pure harmonic waveforms, and in general an RMS measurement is preferred.

PRINCIPLES OF VIBRATION TRANSDUCERS

It is possible to measure the acceleration, velocity, or displacement of a vibration.

Displacement Measurement

To measure displacement, or the distance moved by the measuring position on the vibrating object from its natural position, mechanical, electrical and optical transducers are available. Examples are the vibrograph and the strain gauge.

Vibrograph

An example of the mechanical type of transducer is the vibrograph (Figure 11.12). It is essentially a hand-held box, with a probe connected

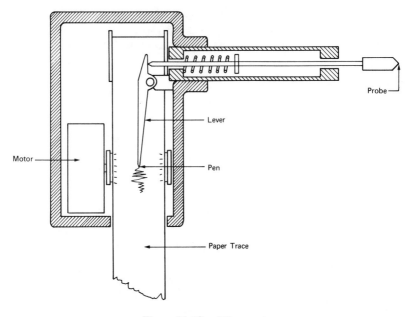

Figure 11.12 Vibrograph.

to a series of levers to magnify the movement and to record the vibration onto a moving trace. The trace can be enlarged optically, and the displacement and frequency measured.

This type is suitable only for low-frequency measurements with reasonably large amplitudes because of the limitations of a mechanical system. High frequency and small amplitudes cannot be read from the trace. Because it is hand-held, some hand movement is inevitable, and this will alter the natural movement of the object. However, as a quick, simple, and cheap method of measuring displacement it is often very useful.

Strain Gauge

The strain gauge (Figure 11.13) can also be used to measure displacement. It is attached directly to the vibrating object, and the strain in the material causes a resistance change that is measured in a Wheatstone bridge. A bridge excitation voltage is supplied and the deformation produces a voltage change proportional to the strain, and hence also proportional to the displacement. The strain gauge is simply connected to a strain indicator of a type suitable for dynamic measurements, such as the type 1526 shown in Figure 11.13.

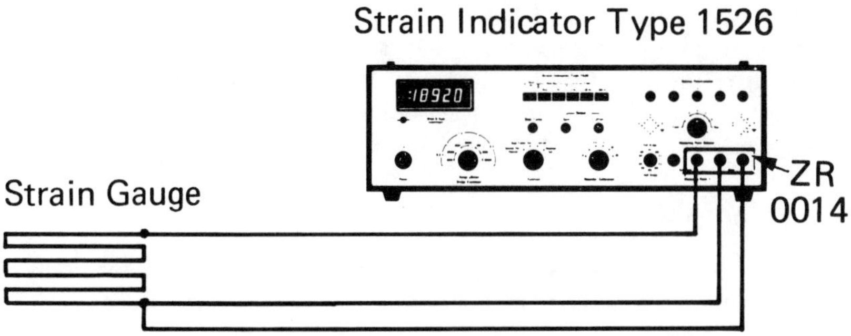

Strain Indicator Type 1526

Figure 11.13 Strain gauge system for displacement measurement.

Velocity Measurement

Velocity, or the speed at which the measuring point on the vibrating object is moving, is measured with a moving element transducer, either a coil or magnet type, which touches the object, or a magnetic one, which is contact free.

Moving Element Transducer

An example of the moving element transducer is shown in Figure 11.14. When the transducer is subjected to vibration the movement of the moving element (in this case the coil) will induce a voltage in the coil. This induced voltage is proportional to the relative velocity of the coil. This

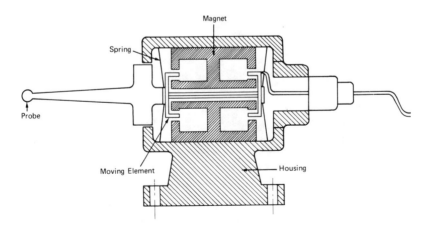

Figure 11.14 Moving element velocity sensor.

sensor is self-generating, and has a low output impedance. However, its moving parts are prone to wear, and it is fairly bulky. It is also delicate and sensitive to magnetic fields and orientation. Small velocity pickups have high internal friction and so are less sensitive. Their directional sensitivity can also change at low frequencies.

Piezoelectric Accelerometer

The piezoelectric accelerometer (Figure 11.15) is the most commonly used transducer for measuring acceleration. It is an electromechanical device that relies on the fact that deformation of a piezoelectric ceramic element by an applied force produces an output voltage proportional to that force. If a mass is attached to the ceramic element and the entire system vibrated, the force needed to accelerate the mass is applied through the ceramic element. Since force = mass x acceleration (Newton's Law), the output voltage is also proportional to the acceleration. The electric charge - voltage x capacitance is also proportional to acceleration, so either voltage or charge may be measured.

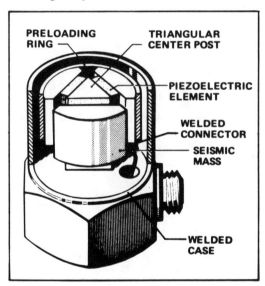

Figure 11.15 Piezoelectric accelerometer.

This type of accelerometer is self-generating with no moving parts to wear, and is very rugged and compact. It has a large dynamic range and a wide frequency range, is relatively inexpensive, very reliable (having high stability), and easy to calibrate. It can be mounted in any orientation to measure the acceleration component along its axis, but it must be rigidly mounted on the test object. Care must be taken to ensure that its mass does not affect the motion of lightweight specimens. Its limitations are that it has a high impedance output and no true DC response.

The frequency response of a typical piezoelectric accelerometer will look similar to Figure 11.16. The lower limiting frequency depends on the cable, the preamplifier, and the environment. The upper limit is set by the method of attaching the accelerometer to the test object.

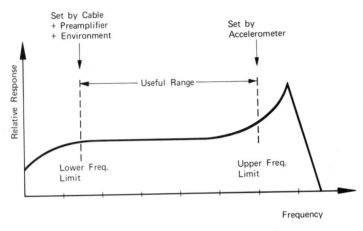

Figure 11.16 Frequency response of a piezoelectric accelerometer.

The responses, with various attachment methods, are shown in Figure 11.17. The best method is a steel stud, although cement, Eastman 910, gives similar results. But magnets and hand probes introduce a spring effect and lower the natural frequency of the system. Measurements are usually made up to about one-third of the resonance frequency. The electrical output of a typical piezoelectric accelerometer is proportional to acceleration over the useful range (Figure 11.18). The lower limit is generally set by preamplifier noise, and the upper limit is set by the accelerometer's structural strength. The upper limit is usually many thousand g's.

VIBRATION METERS

A system for measuring vibration consists essentially of the components shown in Figure 11.19. Considering the signal path through the instrument, the first essential component is a transducer to convert the vibration signal

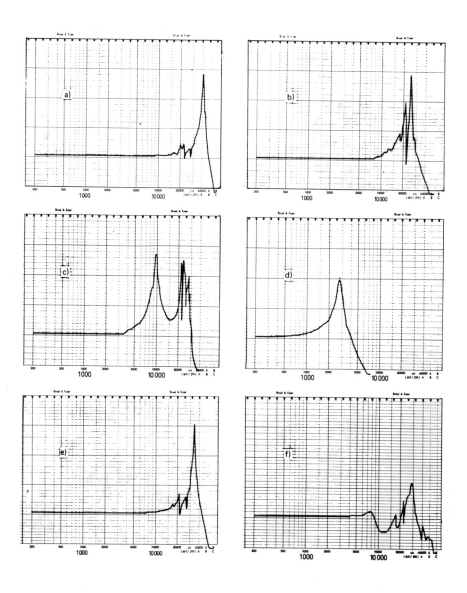

Figure 11.17 Typical frequency response curves of an accelerometer with different mounting methods: (a) steel stud, (b) isolated stud, (c) permanent magnet, (d) hand-held with probe, (e) wax, and (f) soft glue (not recommended).

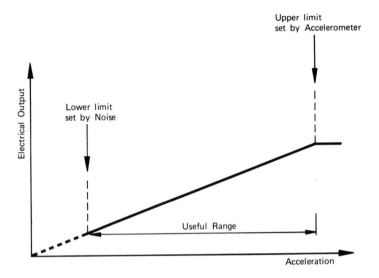

Figure 11.18 Electrical output as a function of acceleration for a piezoelectric accelerometer.

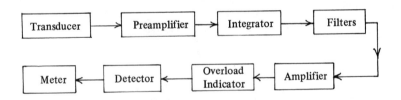

Figure 11.19 Essential components of a vibration system.

to an electrical voltage or charge variation. The preamplifier is required to present the proper electrical load impedance to match the transducer output, thus providing a uniform response of the electrical system across the widest possible frequency range. The integrator is not an absolutely essential component in all cases, but if velocity or displacement measurements are required from an accelerometer input, it becomes essential.

Filters are necessary to limit the frequency range of the instrument to the measurement requirement, and can be provided either internally or externally to the vibration meter. The amplifier provides the dual function of matching the high electrical impedance of the filters and adjusting the signal level to meet the relatively narrow dynamic range of the detector

238 INDUSTRIAL NOISE CONTROL HANDBOOK

circuitry. It is protected by an overload indicator that guarantees that the signal is not distorted from overdriving the amplifier circuits.

The detector determines which of the signal parameters—RMS, peak, or average—is displayed on the meter. Usually it will be the power content of the signal that is of interest, and therefore the RMS value of the signal is needed from the detector. However, the maximum instantaneous level is also of interest in some cases. For example, for peak stress determination the peak value or peak-to-peak function is required from a detector.

An example of a typical vibration meter, designed specifically for the purpose, is shown in Figures 11.20 and 11.21. Tunable filters and graphic recorders are available for making a permanent record for future comparisons. A sound level meter can also be used as a vibration meter, as shown in Figure 11.22. The integrator here is external to the main meter at the input, and a third octave filter set is also in use.

Figure 11.20 Use of vibration meter for routine maintenance inspection of bearings.

FUNDAMENTALS OF VIBRATION 239

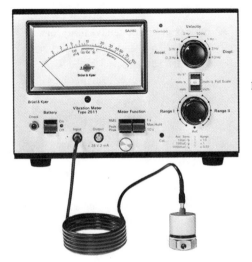

Figure 11.21 Close-up of general purpose vibration meter and accelerometer.

Figure 11.22 A precision sound and vibration meter fitted with octave filter set, integrator adaptor and accelerometer for making vibration analyses.

Applications require that vibration meters and frequency analyzers be portable and operate from internal batteries. They must also be rugged enough to withstand shipment without special care in packing.

CHAPTER 12

VIBRATION CONTROL APPLICATIONS

PADS AND MOUNTS

Large heavy items of motor-driven equipment such as fans, pumps, compressors, punch presses and milling machines are prime examples of equipment that vibrates severely during operation. One method of reducing vibration transmission is to separate each machine, on mounts, so that the shock is contained. Vibration produced is blocked from traveling beyond the machine itself by isolating the equipment so the vibration will not be transmitted to the floor, structural members, or rigid sections of electrical distribution systems.

One of the simplest methods of controlling vibration and shock from machine installations is with isolation pads made of felt, cork, elastomers, steel, aluminum, or numerous combinations of these materials. Pads can be employed by simply placing them under the legs or base of the machine. In textile mills, for example, where there is excessive unidirectional vibration, pads are cemented to the floor and absorb the energy of the low frequency vibrations that are produced. Isolating pads can also be bolted to the floor. Some pad designs require no special attachment method, but are simply put down and remain in place. To prevent movement, the lower surface of the pad may be a waffle-like design that creates a suction with its contact surface. Elastomers are also vibration isolators that work in curtailing detrimental vibrations, which would otherwise be transmitted to floors, foundations and other adjacent machines. These pads will protect a machine from both impact and vibration forces that are transmitted through the floor and foundation from other machines and from vibration caused by the machine itself. Isolators also provide a slight leveling action because of their resilience and their ability to keep the structure free from nonuniform stresses that might be produced by vibration from external sources. Most pads resist petroleum products, cleaning compounds, and deteriorating effects of sunlight and air.

Inertia blocks are also widely used when there is a possibility of shock and vibration forces being transmitted through the foundation. These blocks are concrete sections that are set in the floors or foundations and are isolated from the surrounding material by vinyl-bonded pads, which are inserted before the concrete is poured to act as an impervious barrier. Motors and machines can be mounted on massive inertia pads that are spring-mounted above floor slabs. Construction of pads varies; one of several possibilities consists of a reinforced concrete deck with an outer steel frame incorporating the spring housings.

With the use of relatively light-weight motors and machines, additional isolation can be affected by short rigid sections of electrical conduit stubbed down through floor slabs, passing through an amply sized sleeve. The sleeve is then packed with fiberglass and caulked to make it airtight. The flexible conduit that extends through the lower slab continues to a remote switch gear location. Another method for isolating vibration from rigid electrical distributor systems is to stub-up motor feeder conduit adjacent to, but not in contact with, the inertia pad, then secure a feeder terminal box to the outer steel frame of the floating slab. A large hole is cut in the bottom of the box that surrounds the conduit stub, and the space between the rigid conduit and floating box is later sealed with a thick neoprene gasket.

RUBBER ISOLATORS

Because of economy, ease of installation, and replacement, rubber isolators are much preferred. Some important characteristics of rubber isolators are:

- elastic properties
- damping ability
- ability to reduce noise transmission

Rubber has an excellent damping ability, which can be expressed in the following equation:

$$\phi = \frac{\text{vibrational energy transformed into heat}}{\text{total energy of vibration}}$$

ϕ is the coefficient of damping, and for natural rubber ϕ will average approximately 0.25; for pressed cork ϕ will average approximately 0.12, and for concrete ϕ will average approximately 0.015. The comparison indicates that rubber will change approximately one-fourth of the energy into heat, and is therefore an excellent material to absorb vibration energy.

Extreme temperatures affect natural rubber characteristics, which change its isolation effectiveness. Varying temperatures between 20°F and 140°F will change the characteristics of natural rubber from 12 to 30% for each 36°F change. Extremes in temperature can have permanent detrimental effects on rubber. Rubber isolators are effective when the desirable natural frequency is above approximately 300 vibrations per minute. When natural frequencies below 200 vibrations per minute are encountered spring mountings are usually recommended.

SPRING SYSTEMS

Vibration control can be accomplished by using steel springs in combination with sound-absorbing materials. Because of their effectiveness, springs are widely used in machine foundation design. Advantages of springs are:

- smoother and quieter machine operation
- reduced machine maintenance
- reduction or elimination of the effects of vibration on the structure and occupants
- a decrease in size and mass of the foundation under the equipment
- more accurate design because properties are known and readily predictable
- considered permanent for the life of the machine; properties are not affected by time
- leveling adjustment is possible in spring units
- versatility of installation

Disadvantages of springs are:

- higher initial cost of construction in some installations
- requirements for maintenance of the spring units
- noise transmission and no damping ability

Air Springs for Vibration Control

The ultimate design goal is for vibrationless equipment, but in most cases this is unattainable. Typical problems involve isolating areas that cause undesirable forces. One of the elements available for solving this problem is the air spring. Industrial air suspension systems are used to:

- isolate structures from vibrating masses
- control the frequency of vibrating masses
- keep a suspended platform at a given height under variable load conditions

The air spring is basically a metal container with a rubber liner filled with air. It is a stored energy medium that is designed to support a given load. The important characteristics of any spring in a suspension system are:

1. **Transmissibility**—the ratio of force transmitted through the mounting to the disturbance force. Air springs have low transmissibility coefficients.

244   INDUSTRIAL NOISE CONTROL HANDBOOK

2. Spring rate—the change in load per unit of deflection. The spring rate is directly proportional to the load on the spring and to the pressure of the air in the spring assembly. This is the result of a constant natural frequency suspension, regardless of the load.

3. Load carrying ability—the load range between maximum load and pressure and minimum load and pressure, for which the ratio is 5:1. Air springs have an inherent feature because their load carrying abilities and ratio can be varied by varying their air pressure.

Current designs in air springs are bellows, rolling lobe, bellobe (a combination of bellows and rolling lobes), restrained rolling lobe, filling supported sleeve and hydropneumatic.

Air suspension systems range from simple to complex depending upon the application. If small variations in height are not important and the pressure is constant, then a tire valve can be installed to vary air pressure. The air pressure should be checked periodically and adjusted if necessary. If the same pressure is required in all air springs of a particular system, a pressure regulator valve can be installed to keep the system at constant height. If variable height is desired, automatic leveling valves can be installed. Occasionally it may be necessary to deflate air springs simultaneously so the entire controlled mass rests on solid supports. For this result, a series of check valves may be installed; when the main valve is opened, all the air springs deflate.

An additional advantage of air spring systems is that time has no harmful effect on them. Air columns do not deteriorate with age, nor do they tend to set as mechanical springs might; however, costs of these systems are somewhat higher. Pneumatic systems are easy to install because there is no need for special foundations. Devices are set in place under a machine's support points and connected to plant air supply of 85-100 psi. Readily available units can handle equipment weights from a few hundred pounds to several hundred tons.

ADJUSTABLE DAMPERS

Adjustable dampers operate on the principal that the workpiece can be considered as a part of an entire machine structure and therefore can be the source of chatter. The side opposite the cutting tool contacts the damper with the weakest portion of the workpiece. This absorbs vibratory energy and increases the dynamic rigidity of the machine structure.

The internal resistance of the machine structure is dependent upon a machine structure's dynamic rigidity. Experience has shown that the damping ratio in machine tools is not sufficient to resist resonant vibrations.

When cutting down on speed or the feed rate, productivity has to be sacrificed. With an adjustable damper this problem is eliminated. Other advantages of the external adjustable damper are the cutting process dynamics remain unchanged and the required chatter control is obtained by changing the machine structure.

In this type unit a restraint developed by a piston arm and spring force allows the mass to move up and down. An adjusting screw located on top of the system is used to adjust the amount of spring desired. When the roller is in rotation and the cutting edge passes the roller, the roller is automatically lifted to clear the tool and cut and is then repositioned to place the roller on the new surface. Limit switches, which are located in the lower part of the damping system, control these motions. When these rollers are not in contact with the workpiece, the preload on the spring can be easily retained by a control switch that is located on the piston arm. In the hydraulic type system, there is a flow control valve that restricts the repositioning speed and in turn avoids shock vibrations when the rollers come in contact with the workpiece.

For certain conditions, several sets of springs may be needed to meet various conditions. By using several sets of springs the phase angle, which is dependent upon the damping constant and spring rate, can be carefully controlled. The spring is preloaded to suppress disturbances during the period following the cutting tool's contact with the workpiece. During this period the system approaches a steady state condition.

Coulomb damping is friction between two sliding pads and is often used because it is easy to apply and economical. This type of damping has a quick response. Lubrication is provided between the sliding parts and, as a result, linear viscous damping occurs. There are three screws on each damper to adjust the degree of damping required. In order to prevent any kind of disturbance, friction between the sliding parts should be kept to a minimum; hence the use of lubrication. However, friction can never be entirely eliminated, and a small amount of it is necessary for a system to reach steady state. The dominant factor in the system is stationary; the sliding parts contribute nothing to the system.

The dampers can be located anywhere on a workpiece, but for a lathe they are generally hung on the frame parallel to the centerline of the spindle. Usually on a shorter piece of work only one damper is used, but more can be employed if needed. If several dampers are used on a roll and each one is manually adjusted, a skilled technician should make the adjustments. To avoid this, sensor devices are attached to the dampers and all of the dampers can be adjusted from a center controller. It is important to check the condition of the lubricant; stick-friction should not be used in the system in order to avoid any serious disturbances. The damper can be used in many different applications without any difficulties.

TYPES OF VIBRATION MOUNTS

Cork has been used in many fields of industry as a vibration isolation material. A popular type of cork is made of pure granules compressed together and baked under pressure to achieve a controlled density. Cork materials are used mainly under concrete foundations as illustrated in Figure 12.1. The cork will remain reasonably durable under exposures

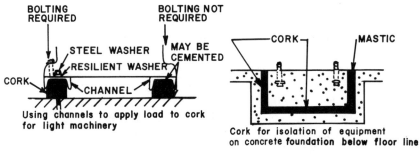

Using channels to apply load to cork for light machinery

Cork for isolation of equipment on concrete foundation below floor line

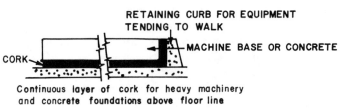

Continuous layer of cork for heavy machinery and concrete foundations above floor line

Figure 12.1 Cork applied under concrete foundations.

to acids, oils, and temperatures between 0°F to 200°F; however, it will be affected upon contact with strong alkaline solutions. Cork will rot, however, from repeated wettings and dryings. As a vibration isolator, cork is limited to frequencies above 1800 cycles per minute. Because of a great degree of damping in cork, the natural frequency cannot be obtained from the static deflection. As an alternative, the natural frequency can be obtained through tests by vibrating the cork under various loads to find the resonance frequency.

Rubber has also been mentioned as a vibration isolator and is useful for frequencies above 1200 cycles per minute. Alkali solutions or acids will not affect rubber, but degradation problems could arise if it is exposed to sunlight. For natural rubber, the temperature range is from 50°F to 150°F; for neoprene from 0°F to 200°F. For applications that expose the rubber to oil, neoprene is more suitable. As rubber ages, it gradually loses its resiliency. The useful life span of a rubber mount is approximately five years under impact applications and seven years under nonimpact applications, although it will retain its sound-insulating properties for much longer. Individual molded rubber mounts are economical only with light- and medium-weight machines because heavier capacity mounts approach the cost of the more efficient steel spring isolators.

A most efficient way to isolate vibration, however, is to use steel spring isolators, which provide greater deflection and thus higher efficiency. Such isolators, illustrated in Figure 12.2, provide deflections up to 1.25 inches. Rubber and other isolators provide maximum deflections on the order of 0.25 inches. Some steel spring isolators can even reach deflections of 10.0 inches. In many other materials costly trial and error tests are usually required, but this is not necessary with steel springs because they closely follow the equations of vibration control.

Steel spring isolators are usually equipped with adjustable snubbers because steel springs contain no damping. Damping is basically helpful in restraining the movement of resiliently mounted machinery but it does reduce the isolation qualities of the mount. Many of the steel spring isolators used by industry have built-in leveling bolts that eliminate the use of shims upon installation. Often, rubber sound-isolation pads are used in conjunction with steel springs because high-frequency noises have a tendency to by-pass steel springs. Table 12.1 illustrates the practical range for various isolation materials at different equipment speeds.

BASIC INSTALLATION PRINCIPLES

When mountings are installed to cut down machinery vibration, the following principles should be observed to ensure maximum efficiency:

1. All parts of the unit of machinery to which the mountings will be applied must be placed on a common rigid base.
2. The entire unit of machinery should be insulated with a suitable mounting.
3. To ensure full effective insulation, elastic connections should be made to the unit as shown in Figure 12.3.
4. If required, grounding should be installed to conduct away any static electricity.
5. If the unit of machinery is tall or unstable, it should be fitted with an enlarged base with a thick, rigid plate.
6. Coupling for elastic force transmission should be done.

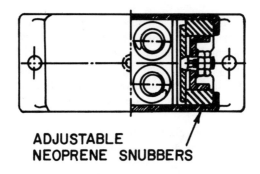

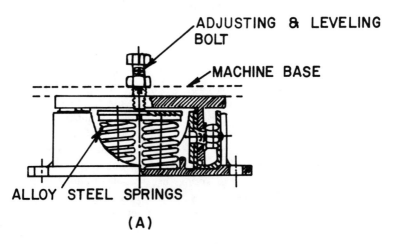

(A)

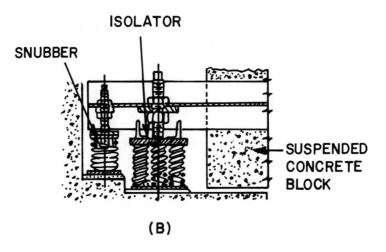

(B)

Figure 12.2 Steel spring isolators.

Table 12.1 Relative Effectiveness of Steel Springs, Rubber and Cork in the Various Speed Ranges

Range	rpm	Springs	Rubber	Cork
Low	Up to 1200	Required	Not recommended except for shock[a]	Unsuitable except for shock[a]
Medium	1200-1800	Excellent	Fair	Not recommended
High	Over 1800	Excellent for critical jobs	Good	Fair to good

[a]For noncritical installations only; otherwise, springs are recommended.

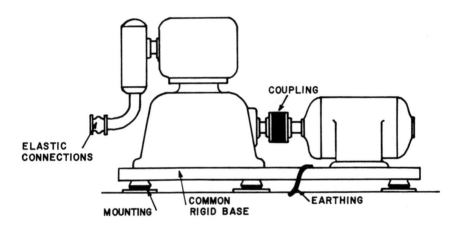

Figure 12.3 Elastic connections are made to ensure full insulation.

SEVERAL TYPES OF INDUSTRIAL MOUNTINGS

Type 1

This mounting, illustrated in Figure 12.4, consists of a cylindrical rubber body with steel plates tightly vulcanized to each end. Each plate has a

250 INDUSTRIAL NOISE CONTROL HANDBOOK

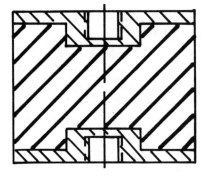

Figure 12.4 Cylindrical steel and rubber type 1 mounting.

threaded center hole for attachment. Upon installation, the top steel plate is bolted to the machine base and bottom steel plate is anchored to the floor. This type of mounting is specifically suitable for vertically loaded units of machinery that only make small sideways motions when in operation. The mounting is flexible and can be made even more so by tilting the insulator; therefore, insulation is achieved through shear and compression rather than just compression. This mounting can be applied to such machinery as converters, fans, pumps and electrical machinery.

Type 2

This mounting, shown in Figure 12.5, is made of two U-shaped steel components with rubber tightly vulcanized between them. There are two holes under the inner shank to improve resiliency qualities. To install, the top U-plate is bolted to the base plate of the unit of machinery and

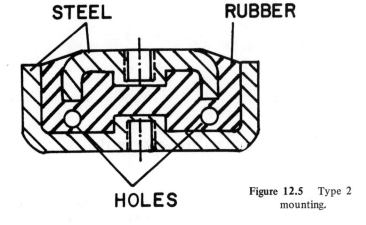

Figure 12.5 Type 2 mounting.

the bottom U-plate is anchored to the floor with an expansion-shell bolt. This type of mounting is especially good for high-speed and heavy machinery because it provides a stable base. It can be applied to machinery such as punches, presses, carpentry machines, weaving machines, and transformers.

Type 3

Figure 12.6 illustrates an antivibration mounting designed to give high resilience and good lateral stability. Mounting is relatively easy since installation height is low. On the bottom of the mounting is a thin layer

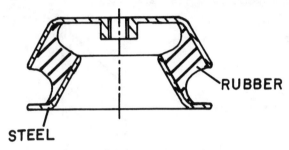

Figure 12.6 Type 3 mounting.

of rubber that will prevent the machinery from moving across the floor. The friction supplied between the mounting and the floor is usually sufficient to make bolting unnecessary. The top metal cover is designed to protect against oil spills. This type of mount is made for both light and heavy low-speed machines, which are generally difficult to insulate, and can be applied to converters, pumps, fans, combustion engines, rolling mills, presses and punches, paper machines and various domestic machines.

Type 4

Another antivibration mounting, illustrated in Figure 12.7, consists of two steel profiles of unequal size with blocks of rubber tightly vulcanized between them. When the rubber undergoes shear stress, the mounting provides resilience. This insulator is used when resilient, soft-mounting is needed for low loads. It is applicable to low-speed machines and light fans as well as various instruments.

Type 5

Another type of mounting is constructed of two identical steel profiles with rubber snugly vulcanized between them (shown in Figure 12.8). It

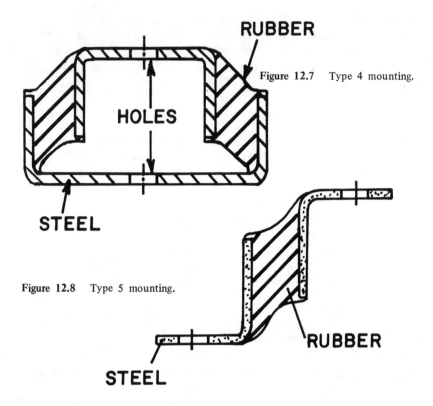

Figure 12.7 Type 4 mounting.

Figure 12.8 Type 5 mounting.

provides simple resilience when the rubber undergoes shear stress. This insulator is applicable when a soft, resilient mounting is needed for low speeds.

Type 6

This type of mounting, illustrated in Figure 12.9, consists of a cylindrical rubber component vulcanized to two steel intermediate plates. Installation of the mounting is not complicated and it can absorb large vertical forces without much deformation. It will also give good stability since its installation height is low.

Providing insulation against horizontal vibrations, it has a high deflection capacity in the horizontal direction, while deflection is relatively low in the vertical direction. In light of these particular deflection properties, we can see that low interference speeds can be horizontally insulated while obtaining a subcritical mounting in the vertical direction with a safe distance

VIBRATION CONTROL APPLICATIONS 253

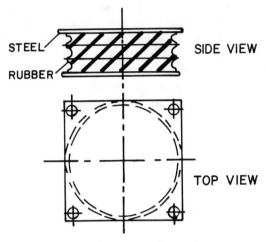

Figure 12.9   Type 6 mounting.

to the resonance point. By connecting two parts in series, large angular movements can be absorbed and a larger vertical resilience can be obtained. This type of mounting is used for such applications as heavy machinery (crushers) and screening plants.

Type 7

The mounting shown in Figure 12.10 provides stability in all directions and is equipped with a leveling screw for varying the height of the unit to the floor. It is installed by first fixing it to the machine and then

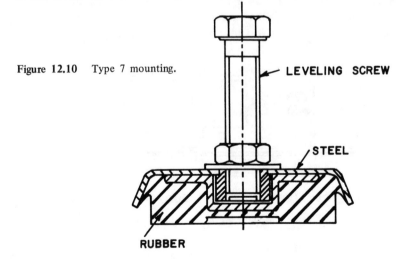

Figure 12.10   Type 7 mounting.

balancing the machine horizontally by using the through-leveling screw. Finally, the level position is anchored with the locking nut. There is a rubber sole at the bottom of the mounting, which provides adequate friction so the machine will not move across the floor.

This type of mounting, which has good lateral stability, is particularly useful for insulating high-frequency vibrations. It is especially applicable to lathes, punches, milling machines, grinding machines, printing machinery and presses.

Type 8

A bumper-type vibration insulator, illustrated in Figure 12.11, consists of a cylindrical rubber component that is snugly vulcanized to a square steel plate. There is a hole at each corner of the metal base plate for

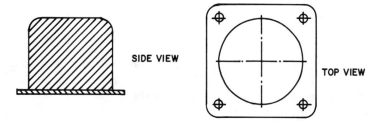

Figure 12.11 Type 8 bumper isolator.

attachment. The rubber component absorbs as much energy as possible and reduces transmitted impact forces. This isolation ensures efficient damping for moving machinery and machine parts. Considerable energy can be absorbed by the resilient rubber bumper. Two bumpers can also be used on a machine to give added resilience (as shown in Figure 12.12).

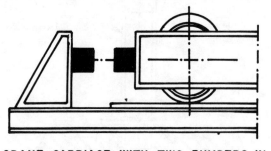

CRANE CARRIAGE WITH TWO BUMPERS IN SERIES

Figure 12.12 Application of the bumper-type isolator.

Two concentric cylindrical metal sleeves with rubber vulcanized between them are designed to take up torsional movements and radial and axial loads. These sleeves can be used to reduce bearing problems in vibrating constructions because the rubber takes the complete movements, so bearing maintenance is not needed. Sleeves that have very good vibration and sound insulating capacity can be applied to machinery such as linkage systems in cars, tractors, railway wagons, bulldozers, agricultural machinery and vibrating screens. A proposed mounting of sleeves is illustrated in Figures 12.13 and 12.14.

Figure 12.13 A proposed mounting of sleeves.

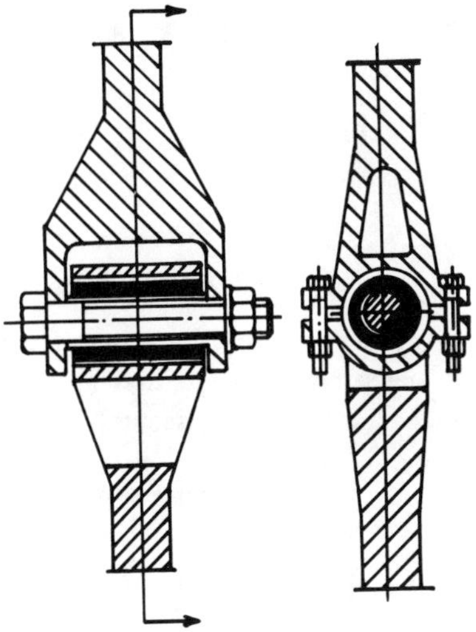

SPLIT BEARING ATTACHMENT

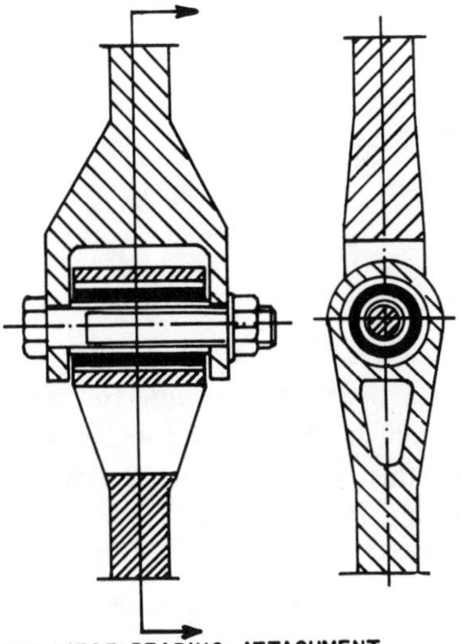

Figure 12.14 A proposed mounting of sleeves.

ONE-PIECE BEARING ATTACHMENT
FOR PRESS FIT TOLERANCE POSITION

SHAFT COUPLINGS

The schematic of a vibration-insulating shaft coupling can be seen in Figure 12.15. It is made up of a U-shaped rubber ring that is tightly vulcanized to two steel rings on either side of it. When in operation, the rubber U-shaped ring will transmit torque from one of the steel rings to the other, thereby delivering the power without vibration.

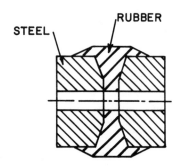

Figure 12.15 A vibration-insulating shaft coupling.

Another type of shaft coupling is shown in Figure 12.16. This coupling has special metal stops that absorb any uncommon torsional stresses. It also has shaft flanges securely screwed to the rubber coupling.

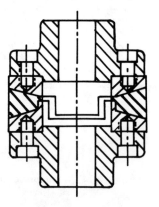

Figure 12.16 Another type of shaft coupling.

In general, shaft couplings protect machinery by providing for smooth, silent and well-balanced running. They can level out any small deviations between the shafts, which could prove valuable in cases where shaft alignment is very difficult. The type illustrated in Figure 12.15 is generally more applicable for smaller machines, and the type illustrated in Figure 12.16 is more suitable for heavier machinery, such as paper machines and excavators, in that it allows for greater power transmission.

VIBRATION PLATES

There are two popular versions of this unit, the single and the double plate. The single plate version has uniform ribs on one of its sides; the double plate version has ribs on both of its sides, which are at angles to each other. These plates, illustrated in Figure 12.17, are made of oil-resistant rubber.

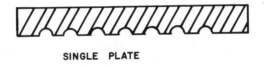

SINGLE PLATE

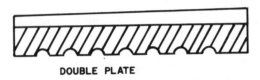

DOUBLE PLATE

Figure 12.17 A typical vibration plate.

These plates are used in industry for relatively simple installation requirements. Best suited for damping a high-frequency vibration, they can be arranged in series to give added cushioning. If mounting is to be permanent, a direct contact between the base and the machinery must be avoided. A mounting example is shown in Figure 12.18.

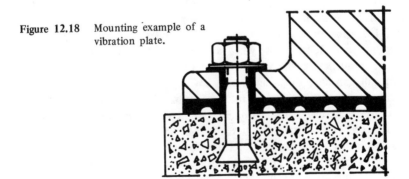

Figure 12.18 Mounting example of a vibration plate.

BLOCKS

This antivibration component is just a simple rubber block with recesses in it for installation (see Figure 12.19). Blocks are easy to fit because

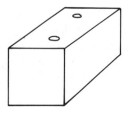

Figure 12.19 A typical antivibration block.

they are placed loosely between the machine and the base. The friction produced between the rubber and the base is sufficient enough to keep the machine from moving. There are four recesses in the block shown, which can be used as sockets for pins or dowels to be positioned to the machine and base.

Blocks can be used to insulate heavier machines with low interfering frequencies. Particularly useful for large composite machines that are on a common girder frame, they are also applicable to many types of machinery including converters, mixers, gear wheels and rolling mills.

APPLICATIONS AND CASE HISTORIES

Some Applications for Isolators

Properly designed mountings allow the installation of some of the heaviest machinery in penthouses, near offices, and many other areas where vibration and noise would be a nuisance. The Federal Housing Administration (FHA) has approved the use of heavy machinery on many penthouse installations provided it is insulated with steel spring isolators. When heavy machinery is installed on upper floors of such complexes, careful consideration must be taken to prevent the transmission of vibration to several floors below when a ceiling, wall, or fixture has the same natural frequency as the vibration. In cases of such resonance vibration the result can be a disturbing noise. When mountings are selected, heavy machinery can be installed in more economically constructed and lighter structures with fewer noise problems. Machinery can also be installed in older buildings that may not have been specifically designed to handle such equipment with the aid of carefully selected mounts. Figures 12.20 and 12.21 show the results of mountings applied to machinery.

In such structures as anechoic room installations, steel spring isolators resting on rubber sound pads are used to stop the transmission of structure-borne vibration and the resultant noise to these rooms. In many air-conditioning installations, the noise and vibration that travel through the piping poses annoying problems. However, if the refrigerating compressors

VIBRATION CONTROL APPLICATIONS 259

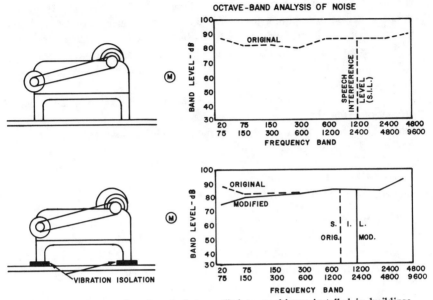

Figure 12.20 Results of mountings applied to machinery installed in buildings that were not constructed to accommodate them.

PUNCH PRESS-TYPE MACHINES ON UPPER FLOOR

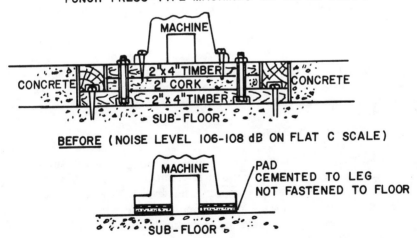

Figure 12.21 Results of mountings applied to machinery installed in buildings that were not constructed to accommodate them.

are installed on mountings, provision should be made for flexibility in the intake and discharge piping to reduce the transmission of vibration. Providing flexibility in the piping itself can be achieved by running the piping a distance equal to 15 times the pipe diameter both horizontally and vertically, and then attaching the piping to the structure. Flexible metallic hose can also be used.

When the piping is suspended from isolation hangers it is given added protection. On the discharge and intake sides of water pumps, flexible rubber hose (approximately three diameters long) should be used. On intake and discharge fans, flexible ductwork should always be used.

Vibration Damping Compound Silences Air-Conditioning System

The James Lick School in San Jose, California, was built with its air-conditioning system fan room directly over an auditorium that was used chiefly for musical presentations. The ducts and related sheet steel fabrication amplified vibrations driven by the pulsating air flow, creating noise levels audible to the students and faculty below.

To prevent this problem, the acoustical consulting engineering firm of Kenward Oliphant specified Korfund Vibrodamper Compound Type 80A because of its high vibration-damping efficiency. The compound dries quickly to a smooth finish like a coat of paint, has a long life and does not become brittle or change in any way with age. Its vibration decay rate in decibels per second is 45 dB at 80°F. Approximately 135 gallons of compound in two coats were applied to the entire exterior surface of the duct system. This application successfully prevented a sound problem. enabling musical performances without disturbing background noise. The air-conditioning system was thus able to function effectively for the benefit of the students and visitors to the school.

Sound Generating Test Chamber

A high-intensity sound testing facility consisting of heavy stiffener reinforced steel exponential horns and reverberation chamber were the materials that needed to be damped. The siren, horn and reverberation chamber systems are housed in a concrete room constructed with 12-inch-thick reinforced concrete walls and a 4-inch-thick steel door. Exponential horns are attached to wide-band noise sirens to increase the efficiency of the sound source and act as an impedance-matching device. The effect of this horn loading on the sound output is considerable, increasing efficiency by approximately 10 dB.

Exponential horns, manufactured by Tenney Engineering, Inc., of Union, New Jersey, although constructed of $3/8$-inch steel plates with $3 \times 3 \times 1/2$-inch angle brackets as stiffeners, cracked and broke from metal fatigue within 10 hours of use when subjected to the 160 to 180 dB noise levels generated by the siren. The continuous flexing of the panels caused the $1/2$ inch bolt heads used in its construction to be sheared off. Horns constructed of a high-test concrete were considered as a possible solution, but the idea was abandoned due to a weight problem. To be technically and economically feasible, a suitable method of damping the horn had to be found.

Korfund Vibrodamper Compound Type 80A was sprayed on the outside of the horn (the inside had to remain smooth to transmit noise) to about one and one-half times the thickness of the $1/4$-inch sheet aluminum (which replaced the $3/8$-inch steel previously used in construction). The first coat was applied to a thickness of about $1/16$ inch, and the following coats were $1/8$ inch thick. A 24-hour drying period was allowed between applications. Since coating the horns with the compound no visible signs of deterioration have been detected. In addition, Vibrodamper was used to coat the siren housing itself as well as the reverberation chamber in which the components are tested. About 100 gallons of the compound were necessary to appropriately coat one horn, siren, and reverberation chamber system of this size. The door was also sprayed with the compound.

The compound permitted $1/4$-inch sheet aluminum with $1 \times 1 \times 1/8$-inch angle bracket stiffeners to be substituted for the $3/8$-inch sheet steel and brackets formerly used, representing a material cost saving of from 25 to 30%. The damping material also permitted reduction of construction time, from 200-300 man-hours to 70-80 man-hours. The use of the compound has resulted in a cost savings of from $1000 to $3000, generally permitting the overall cost reduction by a factor of 3. As a result, the compound is now specified as a standard construction feature in the manufacture of all Tenney exponential horns and reverberation chambers.

Vibration Damping Compound Used in Helicopters

Skin and minor substructure of the aft nacelle area in an S-56 type helicopter were the materials that needed to be damped. In rotating wing aircraft, such as the S-56 (Army H-37 and Marine HR2's), the art of vibration analysis has seen its greatest advancement. Unusual vibration in helicopters has long been recognized as an indication of a need for repair or replacement of some part of the aircraft. Immediate vibration frequency identification is always necessary for accurate diagnosis.

Field service technicians of Sikorsky Aircraft, Stratford, Connecticut, reported continuous fractures of minor structure and skin in the aft nacelle area of the S-56 helicopter. The trouble area was located adjacent to exhaust ejector tubes and the problems were two-fold—vibration and heat.

Previous vibration surveys made in this area indicated a 150-cycle-per-second predominant frequency. The decision was made to use a Korfund Vibrodamper Compound because of its high damping, low weight, resistance to contaminants, and ease of application. Although a survey indicated maximum temperatures below 250°F during warm weather, it was decided that the highest-temperature vibration-damping material available should be used in this area.

The compound 132WS was selected for its high-temperature characteristics and applied in the recommended manner for maximum effectiveness—one and one-half times the thickness of the material to be damped. Test results were gratifying. They showed reductions in vibratory stress in the area of application of from 30 to 50%. On the basis of this test Sikorsky has recommended the use of the compound to the military to alleviate the skin cracking problem.

Shock Isolation Pads Used with Steam Drop Hammer

A 35,000-pound steam drop hammer at one of the facilities of The Steel Improvement and Forge Company of Cleveland, Ohio, was installed on Korfund Heavy-Duty Shock Isolation Pads, without the use of the conventional multilayers of carefully-milled oak timbers and/or pads made of cotton-duck impregnated with neoprene inserted between the anvil and the supporting reinforced concrete block. Before and after the installation, shock tests were conducted by an independent consultant.

The transmission of impact vibrations from a 35,000-pound capacity Chambersburg steam drop hammer had to be reduced from previously experienced levels when standard oak timbers and $5/8$-inch-thick special vibration pads were used. Although vibrations previously produced were well within generally accepted limits, the management of The Steel Improvement and Forge Company wanted to reduce these vibration effects to the lowest practical level, and consulted with Korfund and others for their recommendations on impact isolation.

The isolation problem was complicated by the geological conditions in the area and the magnitude of the weights involved. The foundation rests directly on a rather extensive rock layer passing within variable depths from the surface. The hammer has stationary weight in excess of 1 million pounds, and falling weight of a 35,000 pound ram, plus a die that varies up to half that weight.

To solve the problem within economic reason and achieve improved performance standards, Korfund engineered its new Heavy-Duty Shock Isolation Pads. This particular isolation system specified twelve 1-inch-thick layers of Heavy-Duty Shock Isolation Pads, each layer separated by $1/8$-inch steel plates. This relatively thin stack of isolation pads replaced the two 12-inch layers of timbers, plus three layers of $5/8$-inch-thick cotton-duck impregnated with neoprene. Therefore, a concrete pad was added to the top of the existing foundation to meet the correct operating position of the hammer relative to the floor. No other alterations were made to the existing foundation.

Other types or sizes of hammers and other environmental or geological conditions would necessitate special consideration. Every isolation pad installation is engineered specifically for the requirements of the individual job. The isolation pad installation at the time of hammer erection is simple and rapid, and the installation cost is quite competitive with other conventional types.

The results exceeded expectations. Dr. Edward J. Walter, Director of the Seismological Observatory at John Carrol University, Cleveland, Ohio, showed that in comparison with previously measured results the average value of the acceleration was reduced by a ratio of 6 to 1, or 83%, with a maximum reduction by a ratio of 12 to 1, well in excess of 90%. This tested technique uses a pad system specifically designed for each hammer installation, designed for adequate support and uniform load distribution under both static and dynamic conditions. Of particular interest is the long life feature of the materials used, and the low maintenance cost associated therewith. The Heavy-Duty Shock Isolation Pad system is particularly well suited for replacing conventional isolation materials under installed hammers at time of base removal (see Figure 12.22).

Steel Spring Vibration Isolators

On the basis of neighborhood complaints and disturbances experienced by an electronics facility adjacent to a "town-lot" oil drill rig in the heart of Los Angeles, Korfund Dynamics Corporation was asked to investigate the feasibility of reducing the level of vibration created by the drill rig. Korfund recommendations resulted in the successful reduction of disturbances originated by the drill's operation, as well as curbing the vibration from associated equipment. Effectiveness of the "floating" system engineered and manufactured by Korfund was proven through exhaustive testing (while the rig was in actual use), which indicated scarcely any discernible increase of vibration in the soil.

The Standard Oil Company of California had begun to operate an exploratory oil drilling project in downtown Los Angeles. The facility of

Figure 12.22 Steel spring isolators installed on giant hammer.

of an electronics manufacturer, in which delicate testing and control equipment was operated, was located 120 feet from the drill site. As the oil company began its project, which included the drilling of more than 30,000 feet of hole in several redrills at the same site, the transmission of vibration and shock became intolerable to the electronics firm. Since continuance of the work was of vital importance to Standard Oil, they contacted Korfund as a consultant.

Engineers conducted extensive studies of the drill rig in operation to ascertain the nature of the disturbances generated by the various components. From their analysis, they determined the requirements of resilient mountings (called "vibration isolators") to support adequately each piece of machinery, and to reduce the transmission of vibrational disturbances from the equipment to the neighboring structure through the soil. The derrick itself weighed approximately 176,000 pounds. This weight, plus the more than 200,000 pounds of drill pipe that was suspended from the derrick, totalled more than 376,000 pounds. Eight steel spring isolators supported the entire assembly—two under each derrick leg—to attenuate disturbances originated by the drill's rotation (75-250 rpm) driven through chains, pinions and gears. To include a generous safety factor, each isolator was capable of sustaining a load capacity of more than 121,000 pounds. To simplify the job of leveling-up the installation built-in adjustable leveling devices were included with each isolator.

The draw works and drive mechanism, having a total weight of 104,000 pounds and operated by motors having variable speeds from 0 to 900 rpm, were mounted on 12 steel spring isolators. The capacities of each isolator were varied to accommodate uneven weight distribution. Soil-transmitted-vibration and noise from two Continental-Emsco mud pumps with their driving motors mounted on skids were isolated by placing each skid on 10 isolators per pump. One assembly weighed 74,000 pounds, and the other 52,000 pounds. Each of three General Electric motor-generator sets and compressors were isolated by placing them on 10 isolators. The total weight of each assembly was in excess of 21,000 pounds.

To test the effectiveness of the resilient mounting system, vibration-measuring instruments were placed midway between the operating equipment and the electronics plant. Vibrations were then recorded on a continuous, 24-hour-a-day basis. Even when the drill rig was in full operation, the records indicated that there was scarcely any discernible increase in vibration in the soil over the existing environmental level. As a result of these extensive tests, the vibration isolation system was approved as completely satisfactory.

Vibration Damping Compound with Precision Industrial Scales

Sheet metal housing for industrial scales was the damped material. Precision scales, measuring from 2 to 2000 pounds in 2-pound intervals are used often in factory installations. The Fairbanks, Morse & Co. plant in St. Johnsbury, Vermont, builds an industrial scale for use in areas frequently attended by high noise and vibration. The large sheet metal sides to the scales were affected by vibration and noise from nearby operating machinery, and occasionally resonance effects occurred. Consumer feedback indicated that such vibration would increase the frequency of accuracy checks as well as adjustments to the delicate balancing mechanism. In addition, when loads were applied to scales, clicking noises of levers against stops were audible.

To meet consumer approval and, incidentally, to prolong the already long life of such an instrument, the manufacturer selected Korfund Vibrodamper Compound Type 80A. This was applied by spraying to the inside of scale housing prior to assembly. The application of the compound increased sales appeal by presenting a more solid-sounding housing and by protecting the precision mechanism controlling the accuracy of the product.

ADDITIONAL APPLICATIONS OF SPRING ISOLATORS

1. Steel spring isolators are also used to control vibration due to engine generators as seen in Figure 12.23.

2. Severe vibration in roll grinders is eliminated by installing steel spring isolators as shown in Figure 12.24.

3. Steel spring isolators can protect delicate analytical balances from vibration and disturbances caused by crushing and shaking equipment in large processing plants. Figure 12.25 shows that the highest degree of accuracy is maintained by isolating the concrete block from the table-operator who may then work at the table without disturbing the balance.

4. Figure 12.26 illustrates a 30-foot-diameter rotary hearth furnace at Oldsmobile Division of General Motors Corp., made by the Lithium Company. This furnace "floats" on vibration isolators and is protected from the damaging effects of disturbances created by forging hammers, presses, trucks, trains.

Figure 12.23 Engine generator vibration control. Six of these huge 1425-hp Nordberg engine-generators are mounted on Korfund spring-isolated concrete foundations at plants of the Bangor Hydro-Electric Company, Maine. Isolators permitted installation on poor soil and prevented all vibration transmission.

Figure 12.24 Roll grinder vibration control. 42-inch Lobdell roll grinder, Weyerhaeuser Timber Co., Longview, Washington. Spring-isolated foundation permits precision grinding despite severe vibration from gang saws, trucks, and trains transmitted through water-soaked soil. (Cutaway view shows 2 of 18 vibro-isolators.)

Figure 12.25 Isolators protect an analytical balance against vibration. Steel spring isolators protect this delicate analytical balance from vibration and disturbances caused by crushing and shaking equipment in this large processing plant. Phantom view shows how highest degree of accuracy is maintained by isolating the concrete block from the table-operator who may work at the table without disturbing the balance.

Figure 12.26 Furnace vibration control.

ACKNOWLEDGMENTS

The authors are particularly indebted to the following organizations for furnishing data, illustrations, and permission to use information:

> Trelleborg Rubber Co., Inc., Solon, Ohio 44139, and Trelleborg Gummifabriks AB, Trelleborg 1, Sweden, for data on types of vibration mounts and illustrations.
>
> Korfund Dynamics Corp., Westbury, Long Island, New York 11590, for applications, case histories and illustrations of isolators, spring mounts, vibration dampers.

CHAPTER 13

ABATEMENT AND MEASUREMENT OF CONTROL VALVE NOISE*

INTRODUCTION

Control valves have long been recognized as major sources of noise in fluid process and transmission systems common to the petrochemical and refining industries. Prior to 1968 little work had been done relative to the development of valve noise technology. Since that time, however, significant advancements in valve noise technology have been made. Notable advancements are the ability to predict noise and the development of quiet valves.[1] This chapter deals with techniques for abatement of control valve noise. Use of vibration measurement to calculate level of noise radiated by piping systems is also discussed.

QUIET VALVES

Parameters that determine the level of noise generated by compressible flow through a valve are: geometry of flow path, differential pressure (ΔP) across the restricting elements, ratio of differential pressure to absolute inlet pressure ($\Delta P/P_1$), and the number of ports or restrictions exposed to the flow stream. Figure 13.1 shows four basic approaches to the design of quiet control valves.

Approach (a) utilizes a small tortuous path (diameter $< \frac{1}{32}''$) providing a flow path with a high f ℓ/D ratio designed to maximize the percentage of total pressure drop obtained by viscous stresses induced by the shearing action of the fluid and boundary layer turbulence.

---

*By Ernest E. Allen, Fisher Controls Company, Marshalltown, Iowa.

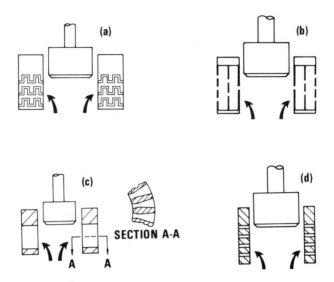

Figure 13.1 Illustrations of valve trim designed for noise attenuation.

The noise characteristic or noise potential of a control valve increases as a function of the square of the differential pressure (ΔP^2) and the ratio of differential pressure to absolute static pressure at the inlet ($\Delta P/P_1$). Thus for high pressure ratio applications ($\Delta P/P_1 > 0.7$) an appreciable reduction in noise can be effected by staging the pressure loss through a series of restrictions to produce the total pressure head loss required. Concept (b) uses multiple series restrictions to limit the $\Delta P/P_1$ ratio across individual restrictions to optimum operating points and provides a favorable velocity distribution in the expansion area. Approaches (a) and (b) can easily be incorporated into cage-style trim fabricated from stacks of discs machined to stage the total pressure drop through a series of circumferential restrictions. It should be noted that both (a) and (b) function as strainers and are more susceptible to plugging as a result of solid particles in the flow stream than single-stage orifical-type restrictions. Multistaging the pressure drop of hydrocarbons can result in the formation of hydrates in the intermediate stages of pressure reduction, resulting in the plugging or blocking of the final stages.[2]

The third approach (c) uses a multiplicity of narrow parallel slots specifically designed to both minimize the turbulence level and provide a favorable velocity distribution in the expansion area. This is an economical approach to quiet valve design and can provide substantial noise reduction (15-20 dB) with little or no decrease in total flow capacity.

It can be shown that the acoustic power of a single flow restriction increases as a function of area squared. Changing the area by a factor of 2 results in a corresponding 6 dB change of power level; in reality, the power level is changed only 3 dB when the number of equal noise sources acting independently is changed by a factor of 2. Critical to the total noise reduction that can be derived from utilization of many small restrictions versus a single or few large restrictions is the proper size and spacing of restrictions so that the noise generated by jet interaction is not greater than the summation of the noise generated by the jets individually. It has been found that the optimum size and spacing are very sensitive to pressure ratio, $\Delta P/P_1$. The author's company has recently developed the technology to utilize the above approach for the design of quiet valve trim using single-stage pressure reduction. This approach, depicted by (d), provides equivalent or better noise performance than (a), (b), or (c).

For control valve applications operating at high pressure ratios ($\Delta P/P_1 > 0.8$), splitting the total pressure drop between the control valve and a fixed restriction (diffuser) downstream of the valve can be very effective in minimizing the noise. To optimize the effectiveness of a diffuser, it must be designed (size and spacing of flow holes in the diffuser) for each given installation so that the noise levels generated by the valve and diffuser are equal. Figure 13.2 depicts a typical valve-plus-diffuser installation.

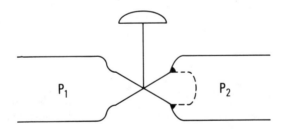

Figure 13.2 Two-stage pressure reduction with a diffuser.

Pertaining to the design of quiet valves for liquid application, the problem resolves itself to one of designing to eliminate cavitation. Service conditions that will produce cavitation can be calculated readily.[3] The use of staged or series reductions provides a very viable solution to cavitation and hence hydrodynamic noise.

PATH TREATMENT

A second approach to noise control is that of path treatment. Sound is transmitted via longitudinal waves through the elastic medium or media that separate the source from the receiver. The speed and efficiency of sound transmission depends on the properties of the medium through which it is propagated. Path treatment consists of regulating the impedance of the transmission path to reduce the acoustic energy that is communicated to the receiver.

The fluid stream is an excellent noise transmission path. When critical flow exists (fluid velocity at the vena contracta is at least at the sonic level), the vena contracta acts as a barrier to the propagation of sound upstream via the fluid. At subcritical flow, however, valve noise can be propagated in the upstream direction almost as efficiently as it is downstream. The impedance to the transmission of noise upstream at subcritical flow is primarily a function of valve geometry. The valve geometry that provides a direct line of sight through the valve (*i.e.*, ball valves and butterfly valves) offers little resistance to noise propagation. Globe-style valves provide approximately 10 dB attenuation. In any path treatment approach to control valve noise abatement, consideration must be given to the amplitude of noise radiated by both the upstream and downstream piping.

Dissipation of acoustic energy by use of acoustical absorbent materials is one of the most effective methods of path treatment. Whenever possible the acoustical material should be located as close to the source as possible. This approach to abatement of valve noise is accommodated by absorption-type inline silencers. Inline silencers effectively dissipate the noise within the fluid stream and attenuate the noise level propagated within the fluid stream. Where high mass flow rates and/or high pressure ratios across the valve exist, inline silencers are often the most realistic and economical approach to noise control. Use of absorption-type inline silencers can provide almost any degree of attenuation desired. However, economical considerations generally limit the insertion loss to approximately 25 dB. Figure 13.3 is a cross-sectional view of a typical inline silencer.

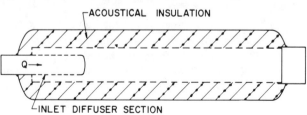

Figure 13.3 Typical inline silencer.

Noise that cannot be eliminated within the flow stream must be eliminated by external treatment or isolation. This approach to the abatement of control valve noise includes the use of heavy-walled piping, acoustical insulation of the exposed solid boundaries of the fluid stream, and use of insulated boxes, rooms and buildings to isolate the noise source.

In closed systems (not vented to atmosphere) any noise produced in the process becomes airborne only by transmission through the solid boundaries that contain the flow stream. The sound field in the contained flow stream forces the solid boundaries to vibrate, in turn causing pressure disturbances in the ambient atmosphere that are propagated as sound to the receiver. Because of the relative mass of most valve bodies the primary surface of noise radiation to the atmosphere is the piping adjacent to the valve. An understanding of the relative noise transmission loss as a function of pipe size and schedule is essential to the development of the most economical approach to noise control of fluid transmission systems.

A detailed analysis of noise transmission loss of pipe is beyond the scope of this chapter. However, it should be recognized that the spectral density of the noise radiated by the pipe has been shaped by the transmission loss characteristic of the pipe and is not that of the noise field within the confined flow stream.

Figure 13.4 depicts the general transmission loss characteristic for commercial piping. The response of pipe to an acoustic field within is characterized by the following frequencies:

1. Coincident frequency, f_c, is defined as that frequency at which the acoustic wave speed is equal to the phase velocity of a flexural wave in the pipe wall.

$$f_c = \frac{\sqrt{3}\, c_i}{\pi\, t c_s} \text{ Hz}$$

2. Acoustic cut-off frequency, f_{co}, is the lower limiting frequency at which acoustic energy propagates freely within the pipe. Below f_{co}, acoustic energy propagates as a plane wave without a radial component to excite the pipe wall.

$$f_{co} = \frac{0.586\, c}{d} \text{ Hz}$$

3. Ring frequency, f_r, is the frequency at which the longitudinal wave in the pipe material equals the pipe circumference.

$$f_r = \frac{c_s}{nd}$$

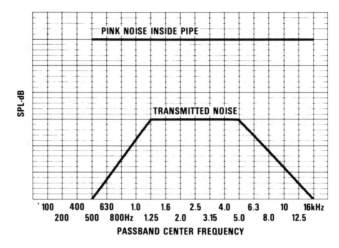

Figure 13.4 Effects of transmission loss characteristic of pipe.

Above the ring frequency, f_r, the pipe wall tends to behave much like a flat plate and mass law applied. Below the greater of the coincident frequency, f_c, of cutoff frequency, f_{co}, the vibration response is stiffness-controlled. In the frequency range between the coincident and ring frequency the resonant modes of pipe vibration are classified as acoustically fast and are relatively efficient in the transmission of sound.

The relative noise transmission loss as a function of pipe size and schedule has been quantified in Table 13.1. For a comprehensive analysis of pipe transmission loss, see Reference 4.

Acoustical insulation of the exposed solid boundaries of the fluid stream is an effective means of noise abatement for localized areas. Test results indicate that ambient noise levels can be attenuated as much as 10 dB per inch of insulation thickness.

Path treatment such as heavy-walled pipe or external acoustical insulation can be a very economical and effective technique for localized noise abatement. However, it should be pointed out that noise is propagated for long distances via the fluid stream and that the effectiveness of the heavy-wall pipe or external insulation terminates where the treatment is terminated.

MEASUREMENT OF NOISE RADIATED BY PIPING SYSTEMS

Control valve noise can be accurately predicted; however, to the best knowledge of the author, reliable techniques for predicting noise resulting

Table 13.1 Correction for Pipe Wall Attenuation

Nominal Pipe Size In.	SCH 10	SCH 20	SCH 30	SCH 40	SCH 60	SCH 80	SCH 100	SCH 120	SCH 140	SCH 160	STD	XS	XXS
1				0		-4.5				-9.5	0	-4.5	-15.5
1½				0		-4.5				-9.5	0	-4.5	-15.5
2				0		-5				-11.5	0	-5	-15.5
3				0		-4.5				-10	0	-4.5	-15
4				0		-5		-8.5		-11.5	0	-5	-15
6				0		-6		-9.5		-13	0	-6	-16
8		+3.5	+2.0	0	-3	-6		-11	-13	-14.5	0	-6	-14
10		+5.0	+2.5	0	-4.5	-6.5		-11.5	-14	-15.5	0	-4.5	
12		+5.5	+1.5	-1	-5.5	-8	-11	-13.5	-15	-17.5	0	-4	
14	+5.5	+2.5	0	-2	-6	-9.5	-12.5	-14.5	-16.5	-18.0	0	-4	
16	+5.5	+2.5	0	-4	-7.5	-11	-13.5	-16	-18.5	-20	0	-4	
18	+5.5	+2.5	-2	-5.5	-9.5	-12.5	-15.5	-17.5	-19.5	-21.5	0	-4	
20	+5.5	0	-4	-6	-10.5	-13.5	-16.5	-19.0	-21.0	-22.5	0	-4	
24	+5.5	0	-5.5	-8	-12.5	-16	-19	-21.5	-23	-25	0	-4	
30	+2.5	-4	-7								0	-4	
36	+2.5	-4	-7	-9							0	-4	
42											0	-4	

from other sources in the system (abrupt expansions, pumps, compressors, pipe fittings) are not available. Thus, actual measurement of the sound radiated to the atmosphere may be required to determine the total contribution made by the piping system to the overall noise level.

A simple sound survey of a given area will establish compliance or noncompliance to the governing noise criterion, but it will not necessarily either identify the primary source of noise or quantify the contribution of individual sources. Frequently piping systems are installed in environments where the background noise due to highly reflective surfaces and other sources of noise in the area makes it impossible to use a sound survey to measure the contribution the piping system makes to the overall ambient noise level.

Vibration measurements provide a viable technique for determining the noise level radiated by a piping system in an environment that precludes the use of sound pressure level measurements.[5] The basic theory employed is the physical relationship between the velocity of a vibrating surface and the acoustic power radiated. Ideally, the pressure of an acoustic wave is proportional to the particle velocity of the medium through which the wave passes, with the constant of proportionality being the acoustic impedance of that medium. At the surface of a pipe, particle velocity is assumed equal to the velocity at which the pipe wall is vibrating. From this, acoustic wave pressure (p_s) at the wall can be related ideally to wall velocity (v) by:

$$p_s = \rho_0 c_0 v$$

where: $\rho_0 c_0$ = acoustic impedance of the atmosphere.

Vibration measurements of the pipe wall can thus be used to calculate sound pressure levels radiated to the atmosphere. The mean squared acoustic pressure (p_s^2) at a point in space a distance r from the axial centerline of the pipe is related to the acoustic impedance of the atmosphere ($\rho_0 c_0$), rms of the pipe surface velocity (v), pipe diameter (d), and radiation efficiency (ζ) as shown below:

$$p_s^2 = (\rho c)^2 v^2 \frac{d}{2r} \zeta$$

The radiation efficiency (ζ) is defined as the ratio of acoustic power at the source to the acoustic power transmitted (W/Wa). The radiation efficiency has been found to equal unity above coincident frequency (f_c) and is directly proportional to frequency below coincidence.

The theory is appropriate for the shell modes only. This type of response would be present if a flat vibrating plate were rolled up into a cylinder. At low frequencies, however, the shell modes are not present and the response of the pipe is due to the entire length of pipe acting as a beam. Beam modes can vibrate with very high amplitudes; however, the efficiency of their coupling to the acoustic field on the outside of the pipe is extremely low. This means that a vibration measurement may indicate high energy content at low frequencies with very little contribution to the observed sound pressure level. Frequencies associated with the lowest shell mode should be considered as a low frequency cut-off for the direct applicability of the theory. This is generally not restrictive in evaluating control valve noise or other broadband and high-frequency noise.

Obviously the accuracy of sound pressure levels (SPL) calculated from vibration measurements is limited to the accuracy of the vibration measurements. The technique is restricted to measurements taken a minimum of two diameters from an end connection of a straight run of pipe. Critical to accurate measurement of pipe vibration is the attachment of the accelerometer to the pipe wall. Ideally, the accelerometer should be rigidly attached to a small metal pad that is welded to the pipe. Also, some device should be used to electrically isolate the accelerometer from the pipe, such as an insulated stud or washer. This attachment method will yield valid information over the entire frequency range for which the particular probe is specified.

An alternative method is to attach pads or studs to the pipe wall, using an adhesive. As long as a stiff, thin-layered adhesive is used, this method can be effective over the specified range of probe. Different adhesives are necessary depending on the temperature of the application. Magnetic attachments should be of special design to give a firm bond to a cylindrical surface. Even with a good magnetic attachment, the high-frequency response is limited. If a magnetic base is used, then the surface should be clean of paint and dirt to ensure maximum contact. Hand-held accelerometers generally are limited to very low-frequency measurements.

It should be recognized that vibration measurements are not a panacea for all of the problems associated with noise analysis of a fluid transmission system. Sound and/or vibration surveys can quantify the noise levels radiated by a fluid transmission system; however, because the noise generated within the flow stream is both structure-borne and fluid-borne, the primary source of noise generation is not necessarily obvious, thereby making it difficult to know which sources (if any) are controlling the sound field.

CLOSURE

From the foregoing discussion it should be obvious that a substantial amount of progress in the area of control valve noise technology has been achieved in the last very few years. Among the most notable advancements are the ability to predict accurately the level and spectral density of control valve noise radiated to the atmosphere via the adjacent piping and the development of quiet valve trim.

From a purely noise consideration, there are few control valve installations that can be considered truly standard. They will be unique from the standpoint of installation geometry, service conditions, noise attenuation requirements, or some combination of these. With so many possible installation variables and the numerous pieces of control valve noise abatement equipment, it becomes extremely important that knowledgable persons are consulted in the application of this equipment. For example, several approaches may be taken to the same problem. One approach might produce the very quietest installation but at a prohibitive cost, whereas another approach could meet the required noise specification at a substantial saving. Without the ability to predict noise levels and without the choices of equipment, optimizing a given installation from the noise and cost standpoints would not be possible.

Where do we go from here? Comparing current noise technology with other important control valve technologies, such as systems analysis and valve sizing, indicates that the noise technology is in an emerging status. If this is the case, then dramatic progress can be expected in the immediate future. Studies in progress are intended to increase the understanding of the noise generation mechanisms and identify parameters not presently being considered. These studies and others should generate new and more efficient items of equipment and result in more precise techniques for the prediction of control valve noise.

REFERENCES

1. Fisher Controls Company, Section 3, Catalog 10, Marshalltown, Iowa 50158.
2. Schuder, C. B. "Thermodynamics of Flow-Through Control Valves," *Instrumentation Technol.* (June 1973).
3. Stiles, G. F. "Development of a Valve-Sizing Relationship for Flashing and Cavitating Flow," ISA Final Control Elements Symposium, Wilmington, Delaware (1970).
4. Fagerlund, A. C. "Transmission of Sound Through a Cylindrical Pipe Wall," ASME November 1973, Paper No. 73-WA/PID-4.
5. Fagerlund, A. C. "Conversion of Vibration Measurements to Sound Pressure Levels," TM-33, Fisher Controls Company, Marshalltown, Iowa 50158.

CHAPTER 14

HYDRODYNAMIC CONTROL OF VALVE NOISE*

Liquid flow through a control valve (called hydrodynamic flow) can and often does create noise. There are three categories of hydrodynamic noise: noise from noncavitating liquid flow, cavitating liquid flow, and flashing liquid flow. Of the three, cavitating flow is the major noise problem. Laboratory testing and field experience show that noncavitating and flashing liquid flow noise levels are quite low and generally not a problem.

Cavitation is a two-stage phenomenon. The first stage involves the formation of vapor bubbles in the fluid stream. As liquid flow passes through the orifice of a control valve, its velocity causes pressure at the vena contracta to drop below the vapor pressure of the liquid, and vapor bubbles are formed. The vena contracta is the point beyond an orifice where the flow cross-section is smallest, pressure is lowest and velocity is highest. However, since clearance between the valve plug and seat ring is the primary restriction in most conventional valves, the vena contracta is formed near the valve seat line.

The second stage is the implosion of these vapor bubbles. As the fluid moves downstream from the vena contracta into a large flow area, velocity decreases with a resulting pressure recovery. When static pressure exceeds the vapor pressure of the liquid, the vapor bubbles implode, generating extremely high-pressure shock waves that hammer against the valve outlet and piping. (Pressures in the collapsing cavities reportedly can approach 100,000 psi in magnitude.) Noise and damage result.

---

*By A. C. Casciato, Fisher Controls Company, Marshalltown, Iowa.

CONTROLLING CAVITATION

Solving cavitation problems begins first with either controlling the cavitation process or, ideally, eliminating cavitation altogether. In controlling cavitation, the techniques employed are often defensive in nature. For example, valve parts subject to damage are furnished in hardened materials in an attempt to extend valve life against the erosion and shock generated by the imploding vapor bubbles. Another technique simply lets cavitation exhaust itself by destroying some sacrificial part of the piping system (an elbow downstream of the valve orifice plate).

A third technique, developed within the past few years, involves special cage-type valve trim for globe valves that moves the primary fluid restriction away from the valve plug seat line. In new trim design, a number of pairs of small, diametrically opposed flow holes are located in the wall of the cage, Figure 14.1. As the valve plug moves away from the seat, increasing numbers of these holes (always in pairs) are opened to the inside of the cage. Each hole admits a jet of cavitating liquid that ensures

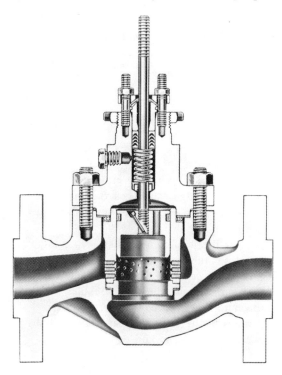

Figure 14.1 New technique in trim design where a number of pairs of small, diametrically opposed flow holes are located in the cage wall.

substantial pressure recovery in the center of the cage. Collision of this jet with that of the opposing hole creates a continuous fluid cushion, Figure 14.2. The cushion in turn prevents cavitating liquid from contacting the valve plug and seat line of the valve. Under certain conditions, this trim can reduce valve noise as much as 6 dB.

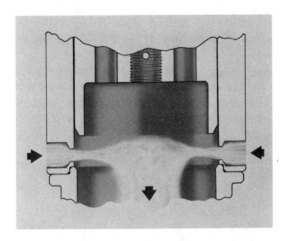

Figure 14.2 Holes shown in Figure 14.1 design admit a jet of cavitating liquid. Collision of this jet with that of opposing hole creates a continuous fluid cushion.

Ball valves are often used in installations that require wide ranges of flow control. A ball valve, however, is a device having a greater pressure recovery from the vena contracta pressure than that given by a globe valve. This pressure recovery advantage has its price in that a cavitating condition is more easily reached. An answer to this problem is a special trim consisting of a bundle of parallel flow tubes that, under certain conditions, create a back pressure within the ball valve to keep static pressure above the vapor pressure of the fluid. However, when cavitation does occur, these flow tubes restrict the size and number of vapor bubbles. Damage tests with soft aluminum rods indicated that damage downstream of the flow tubes was insignificant as compared to damage without them.

When cavitation control methods are used, noise reduction is gained by applying acoustical insulation on the valve and associated piping, by using heavy-walled piping, by installing the valve in an enclosure, or by burying the pipeline. These techniques, commonly referred to as path treatment, do not reduce the level of noise carried in the fluid stream; they only shroud it. Therefore, it is important to note that where path treatment stops, fluid noise may annoyingly reappear.

ELIMINATING CAVITATION NOISE

If the cavitation is eliminated, cavitation-created noise is also eliminated. Several techniques can be applied to eliminate cavitation. One involves placing the control valve within the piping system at a point where pressure drop and fluid temperature conditions will not create cavitation. If this proves impossible, then two or more valves in series, each taking a portion of the total desired pressure reduction (called staging), can be used to prevent dropping pressure within the valve below the vapor pressure of the fluid.

The technique of staging has been designed into special valve trim for globe bodies. An example of this style trim is a cage consisting of one or more concentric, cylindrical sections that contain specially drilled orifices (Figures 14.3 and 14.4). In operation, each section stages the pressure

Figure 14.3 Cage trim with drilled orifices is designed to eliminate cavitation at pressure drops to 3000 psi.

drop, the number of stages required depending on the inlet pressure and the total pressure drop across the valve. This type of trim would be applied normally when the pressure drop is in the 1000 to 3000 psi range, and it can be characterized when the pressure drop decreases with increasing valve travel. The characterization provides pressure staging and cavitation prevention at low valve plug lift, with the staging effect becoming progressively less as the pressure drop becomes smaller at greater valve travel.

A problem is encountered in this trim design as the pressure drop exceeds 3000 psi. With the valve plug off its seat, the pressure drop-staging devices below the plug seat line are open to flow and perform as designed.

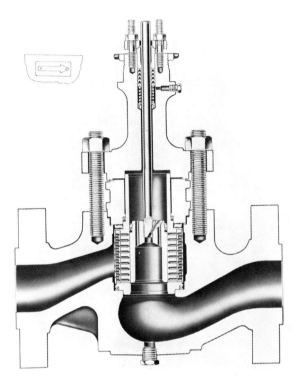

Figure 14.4

However, pressure drop staging does not occur in those devices still blocked to flow; as a result, full inlet pressure registers against the valve plug through them. Due to plug-cage clearances, flow at nearly full inlet pressure moves between the plug and cage into the downstream pressure area. This high-pressure, high-velocity flow stream exiting into the downstream flow passage may cavitate near the valve plug seat line. The higher the inlet pressure, the greater the possibility of damage.

This clearance flow problem is solved by a newly-developed trim that stages both normal flowing pressure drop and all clearance flow between the valve plug and cage (Figure 14.5).

PROBLEM DEFINITION

To optimize the noise control effort, it is essential first to determine existing or potential operating noise levels. Through research, a hydrodynamic

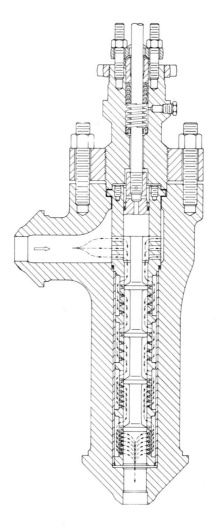

Figure 14.5 High-pressure clearance flow problems are solved by this new trim configuration.

noise prediction technique has been developed that involves the factors of valve style; size and type of trim, size and schedule of adjacent piping; inlet pressure, pressure drop, and liquid vapor pressure; and valve capacity.

$$SPL = SPL_{\Delta p} + \Delta SPL_{c_v} + \Delta SPL_{\Delta p/(p_1-p_v)} + \Delta SPL_k$$

where: SPL = overall noise level in A-weighted decibels (dBA) at a predetermined point (48 inches downstream of the valve outlet and 29 inches from the pipe surface).

$SPL_{\Delta p}$ = base SPL in dBA, determined as a function of pressure drop (ΔP)

ΔSPL_{C_v} = correction in dBA for required liquid sizing coefficient (C_v)

$\Delta SPL_{\Delta p/(p_1-p_v)}$ = correction in dBA for valve style and pressure drop ratio $\Delta p/(p_1-p_v)$. The pressure drop ratio must be calculated using inlet pressure p_1 (in psia) minus the vapor pressure of the liquid (p_v in psia).

ΔSPL_k = correction in dBA for acoustical treatment, such as heavy wall pipe or insulation.

The predetermined point for SPL was selected because of the physical dimensions of a "soft room" (a sound absorbing chamber) that was placed around test valves in the laboratory. The point of measurement inside this room was 48 inches downstream of the valve and 29 inches from the pipe surface. There are no control valve industry measurement standards with regard to distance from the noise source; however, since noise attenuates with distance, some reference point always should be selected and recorded.

Use of this prediction technique provides a valid and ready determination of a valve's noise level. The noise level then becomes the basis for evaluating available noise treatment methods. The following example illustrates use of this technique.

Given: Valve style, size and type of trim—a 2 inch, ANSI Class 300-pound rated Fisher Design ED with standard trim.

Adjacent piping: 2-inch Schedule 40 pipe
Inlet pressure (p_1): 250 psi
Pressure drop (ΔP): 175 psi
Calculated required C_v: 70
Flowing medium: Water at 200°F
Vapor pressure (p_v) of water at 200°F: 11.5 psia
Solution: Calculate the pressure drop ratio

$$\frac{\Delta p}{p_1-p_v} = \frac{175}{250-11.5} = 0.734$$

From Figure 14.6, $SPL_{\Delta p}$ = 59 dBA. From Figure 14.7, ΔSPL_{C_v} = 37 dBA. From Figure 14.8, $\Delta SPL_{\Delta p/(p_1-p_v)}$ = -8 dBA. From Table 14.1, ΔSPL_k = 0 dBA.

$$SPL = SPL_{\Delta p} + \Delta SPL_{c_v} + \Delta SPL_{\Delta p/(p_1-p_v)} + \Delta SPL_k$$

$$= [59 + 37 + (-8) + 0] \text{ dBA}$$

$$= 88 \text{ dBA}.$$

290 INDUSTRIAL NOISE CONTROL HANDBOOK

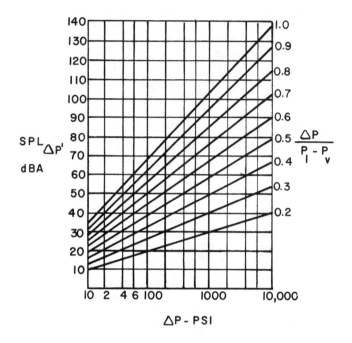

Figure 14.6 Base $SPL_{\Delta p}$ for all valve styles.

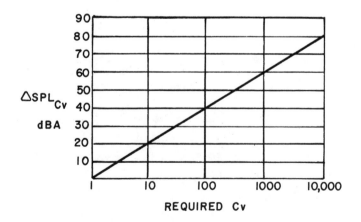

Figure 14.7 ΔSPL_{C_v} correction for all valve styles.

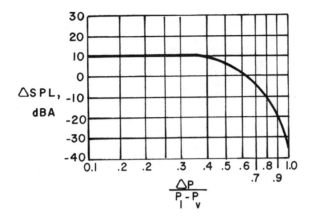

Figure 14.8 $\Delta SPL_{\Delta p/(p_1-p_v)}$ correction for standard cage trim in a globe valve, flow down.

Table 14.1 ΔSPL_k Correction for Pipe Wall Attenuation (dBA)

Nominal Pipe Size (in.)	Pipe Schedule Number							
	30	40	80	120	160	STD	XS	XXS
2 2	–	-19.3	-23.7	–	-29.1	-19.3	-23.7	-31.9
4	–	-22.5	-26.9	-30.2	-32.5	-22.5	-26.9	-35.4
6	–	-24.5	-30.0	-33.3	-36.3	-24.5	-30.0	-38.6
8	-23.9	-25.9	-31.5	-36.0	-38.9	-25.9	-31.5	-38.5
10	-24.8	-27.0	-33.2	-37.7	-41.3	-27.0	-31.1	–
12	-26.3	-29.0	-35.7	-40.5	-43.8	-28.0	-31.7	–

Should the predicted noise level exceed established noise standards, federal or other, a choice must be made whether to: (1) control cavitation and use path treatment methods, or (2) prevent cavitation. At first glance, the simple answer is to prevent cavitation, but in actuality, the decision is based on economic trade-offs. Table 14.2 provides a general comparison of noise control effectiveness and economics for the various techniques.

Table 14.2 Hydrodynamic Valve Noise Control Techniques

Technique	Effectiveness	Economic Aspects
Hardened trim parts, sacrificial piping	Provides no noise attenuation. Must rely on path treatment.	Frequent need for replacement valve parts and new piping proves very expensive.
Valves in series	Depends on service conditions. This technique can eliminate cavitation and noise without requiring additional treatment methods.	Large capital investment in valving plus the cost of designing and installing complex piping configurations may make this method impractical.
Acoustical insulation, heavy walled piping	Insulation can provide up to 10 dB attenuation per inch of thickness. See Table 14.1 for pipe wall attenuation.	Very economical in comparison to special valves and special piping configurations but cavitation damage will still result unless preventive steps are taken.
Cage-style trim for cavitational control	Highly effective in controlling cavitation up to 1440 psi pressure drop. Provides some noise attenuation.	Within its limitations, this technique generally proves to be more economical than valves in series or valve trim designed to prevent cavitation.
Valve trim to prevent cavitation by staging pressure drop.	Effective at most pressure drops. Limited in capacity and restricted to use on clean liquids.	High initial cost, but usually the most economical approach when pressure drops exceed 1440 psi and capacity requirements are not extreme.

CHAPTER 15

VENTILATING SYSTEM NOISE CONTROL*

INTRODUCTION

Air-conditioning systems vary greatly in design and purpose. For example, for heating a building, a heat-pump must be included in the design. When humidity control is desired, reheating coils following an evaporator cooling system are included. In industrial sites or hospitals requiring air sanitation, electric precipitators, wet collectors, or ultrasonic agglomerators may be necessary. These designs may also require air sterilizers and activated carbon absorbers for odor removal. When only ventilation is necessary, a forced-air draft system is employed whereby a fan directs filtered air to designated locations in the building.

Any of the many components of an air-conditioning system may be a cause of unwanted noise generation or vibrations. Adequate noise control of these systems can be achieved by careful examination of the possible causes and by providing proper isolation and insulation to the primary units in the system.

NOISE SOURCES

Ventilating Systems

In general, ventilating systems consist of a motor-driven blower that directs air through a plenum into headers for distribution to various parts of a building. Often the fan is located in the plenum to isolate noise.

Air flow is regulated by the difference in pressure between the two terminals of the duct network. Designs are classified according to low,

---

*By N. P. Cheremisinoff and E. J. Bonano, Clarkson College of Technology, Potsdam, New York 13676.

medium, and high-pressure systems. The maximum allowable static pressure at any supply or return opening for each type is given in Table 15.1.

Table 15.1 Classification of Ventilating Systems

Type System	Maximum Static Pressure (inches of water)
Low pressure	0.5
Medium pressure	3.5
High pressure	6.0

Two principal types of noise are normally encountered in simple ventilating units—fan noise and air flow noise. Fan noise can be subdivided into two components—rotational and vortex. The rotational component is associated with the impulse transmitted to the air each time a blade passes a fixed position. Hence, it is a series of discrete tones at the fundamental blade-passing frequency and is a function of the harmonics. The vortex component of noise is attributed largely to vortices shedding from the fan blades. This occurs due to air turbulence caused by the wind stream passing over system components such as elbows or filters. Medium and high pressure systems are often plagued with turbulency noise generation.

For short ducts, noise generation is derived primarily from the fan. However, for long distribution systems, in addition to air turbulence, other sources may exist. For example, fan noise may become masked by sounds of air turbulence and radiations from entrance ports. Wall and panel vibrations originating from the operating machinery may also exist and might result from equipment failure such as fan unbalance or bearings. In addition, brush or magnetic noise may develop in the fan.

Air-Conditioning Systems

The location of possible noise trouble spots becomes complicated with more elaborate systems because there are a large number of components that could be the cause. The major units that generally require attention are compressors, condensers, cooling towers, evaporators and pumps and piping.

Compressors

Compressors must be located relatively far from the distribution system. They are often mounted on an elaborate mounting block on the top floor of a building. To provide noise control, a floating concrete floor is constructed with a suspended ceiling with acoustical paneling directly beneath it.

Condensers

The two types of heat exchangers, free and forced-convection, are noise radiators that cannot be readily isolated in a special housing, as maximum surface exposure to air is necessary for high heat-transfer efficiency.

Cooling Towers

These units, generally having several hundred cubic feet of volume, are used for cooling water that has been used in heat exchangers that remove heat from the compressed refrigerant. The heat is removed by partial evaporation into the atmosphere. The process is enhanced by tube-axial or centrifugal fans, belt-driven by hermetically sealed motors.

These can be extremely noisy units with severe vibration problems. They are generally mounted on the building's roof with a parapet often placed around them to reduce airborne noise. Vibration pads must be included as part of the mounting design.

Evaporators

Noise is produced by these units because the two-phase, liquid-gas mixture exiting the capillary causes a high-pitched whistle.

Pumps and Piping

Piping from pumps can undergo severe vibrations or cavitation, particularly when not properly suspended.

NOISE CRITERIA

Federal agencies and organizations are establishing detailed criteria for acceptable noise levels. Different parameters and values have been used by them. Unfortunately, many different rating procedures have evolved for estimating noise levels. The more common constructs and terms related to ventilating system noise will be discussed.

Loudness and Loudness Level

Loudness level is the logarithm of the quantity loudness. Both are a means of rating noise based on a person's judgment. The procedure for calculating loudness rates and relative loudness is standardized both in the U.S. and internationally.

Speech Interference Level

A calculation procedure that rates steady noise according to its interference with communication is referred to as the speech interference level.

NC and PNC Ratings

Noise Criteria (NC) curves are used extensively in rating noises in buildings and in the rating of air-conditioning noises in particular. Preferred noise criteria (PNC) curves are essentially a modified NC plot.

Figure 15.1 shows the NC curves. NC values apply to steady noises and specify the maximum noise levels permitted in each octave band for a specified NC curve. NC curves have been used cautiously in the past

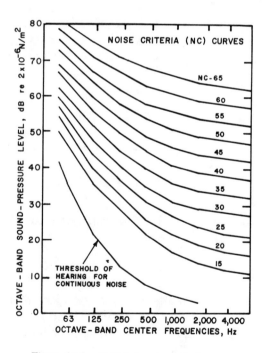

Figure 15.1 Updated noise criteria curves.

because they tend to underestimate actual sound pressure levels. This is particularly true for ventilating systems in which fan or duct turbulence is a major problem. To design for an acceptable noise quality, it is common to lower levels 5 dB in both the very low- and high-frequency bands.

PNC curves (Figure 15.2) have values approximately 1 dB below the NC values in four-octave bands, to compensate for the NC inaccuracy (at 125, 250, 500 and 1000 Hz for the same curve ratings). In the highest three bands and the 63 Hz band, values are 4 to 5 dB lower.

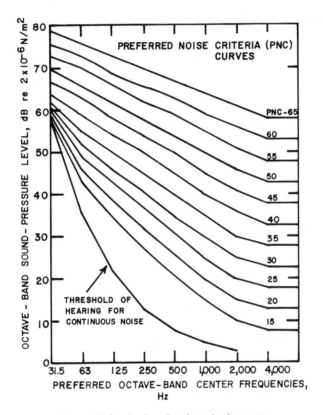

Figure 15.2 Preferred noise criteria curves.

Noise Pollution Level (NPL)

NPL is a procedure for evaluating noises from several sources and fluctuating or intermittent sounds. The noise pollution level (L_{np}) utilizes data (either A-weighted noise levels or loudness level) that are recorded over a sufficient length of time to establish the statistical nature of the total noise exposure. The noise pollution level is expressed as:

$$L_{np} = L_{eq} = 2.5 \, \sigma \text{ dB (NP)} \tag{15.1}$$

where: L_{eq} = mean-square, A-weighted sound level over some sample time, dBA
σ = standard deviation for the A-weighted sound level data

A-Weighted Sound Level (L_A)

This sound level is a means of correlating speech-interference level and NC or PNC levels. This parameter is directly measured by a sound-level meter that has an electronically modified frequency response referred to as A-weighting. The units are in decibels, and generally an A is denoted to signify the weighting scale (*i.e.*, dBA).

Insertion Loss

This term is used to denote the noise-level reduction as measured at a specified point in a channel after a duct silencer is positioned ahead of the measuring station.

Transmission Loss (L_{TL})

Referring to the acoustical signal reduction of a duct silencer, transmission loss is measured through a standard test by positioning a loudspeaker at the entrance of the duct network, and a microphone at an exit station. Tests are made with no air flowing through the system.

Dynamic Transmission Loss

Referring to the signal reduction of a duct silencer as air passes through the channel, this parameter is generally much less than transmission loss because of air turbulence.

Different methods have been adopted to account for the effect of noise in a working environment, with recommended exposure levels varying. Standardization is not yet attained in noise evaluation. Criteria depend not only on the method of evaluation but on the specific environment; therefore, a range of ratings is usually indicated. For areas requiring a high-quality environment, the lower ranges are recommended.

Where economy or physical limitations exist, the upper edge should be employed.

When employing PNC or NC curves, the sound-pressure level in all octave frequency bands should not exceed levels denoted by the appropriate curves. In practice, the noise level in one octave band is permitted to exceed the corresponding value on the specified criterion curve by no more than 2 dB.

The U.S. Department of Housing and Urban Development has established fairly extensive guidelines for noise exposure levels. The interior noise exposure in new and rehabilitated residential construction should not exceed 45 dBA for more than an accumulation of 8 hours in any 24-hour day. Guidelines for speech interference levels are also outlined. Figure 15.3 illustrates these recommendations.

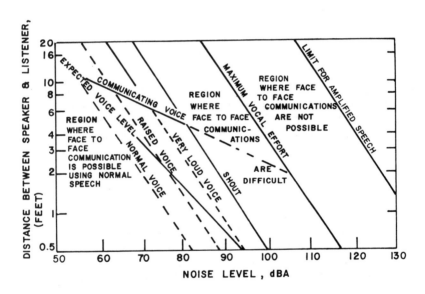

Figure 15.3 Graphic guide to allowable noise exposure levels.

MEASUREMENT TECHNIQUES

The sound-level meter consisting of a microphone, amplifier, calibrated attenuator, a series of frequency-response shaping networks, and an indicating meter is most often used to measure noise levels. Weighting scales in American instruments are usually denoted A, B, C, and Flat.

The C network is used to limit the low and high-frequency response of the device so that the instrument will not respond easily to signals outside the audible frequency range. A-weighted readings are used to estimate the probability of hearing damage in industry. These values have also been correlated with the annoyance caused by traffic and aircraft noise. The B-weighting network is rarely used. The Flat response of the meter is employed as a calibrated input to a frequency-band filter set or tape recorder.

The sound-level meter weighting networks were chosen by the American Standards Institute as the reference curves for sound loudness contours for the three frequency responses available in the instrument. In general, these devices are not well suited for measuring intermittent noises that vary rapidly with time. In such cases, high-speed level recorders (graphic level recorders with high-speed pens) are employed. Such devices can record transient sounds whose sound-pressure level rises 25 dB within 25 msec.

Analyzers are devices that provide information on the distribution of the noise signal energy as a function of its frequency. For continuous noises, the meter's signal is first recorded in the field on a magnetic tape. The noise is then reproduced and analyzed in the laboratory. Analysis with octave filters provides the necessary information for acceptable levels.

A sound generator is used for measuring sound-transmission loss, absorption, and reverberation time. These parameters are evaluated in terms of discrete frequency bands or sinusoidal frequencies. Oscillators or random-noise generators are the most commonly used signal sources.

NOISE REDUCTION

Noise transmission is greatest in unlined metal ducts. The amount of noise-reduction occurring in such designs is almost negligible (on the order of 0.1 dB of sound reduction per foot of duct even at frequencies greater than 1000 Hz). Whenever economically feasible, sound-absorbent materials should be introduced in a duct as lining. Sound-attenuation calculations become complicated as a large number of physical data become necessary. Information on acoustic resistance, acoustic reactance, and the phase angle between the two are required. Empirical correlations exist and calculation procedures have been developed for determining noise-level reductions in ducts lined with sound-absorbent materials.[2,7]

It is often good design to install a traverse baffle or a sharp-angled bend in the channel. As sound travels down a lined duct, it becomes more highly absorbed around the perimeter than in the center. This results in a sound-pressure loss at the duct boundaries caused by destructive interference between the boundary and direct-reflected sound at grazing incidence. A sound-absorbent baffle can reduce the sound energy concentration near the center of the duct by causing both absorption and reflection.

The use of package attenuators has proven successful in air-conditioning noise reduction. Attenuators are short duct inserts whose cross section is comprised of zigzag paths. These paths achieve high acoustic absorption without causing large frictional resistance to airflow. Many designs are available commercially.

In designs requiring a large number of ducts that are supplied by one main fan, plenum chambers should be employed. A plenum chamber is a muffler, the interior of which is lined with sound-absorbent material. Figure 15.4 illustrates a single plenum chamber (note that inlet and exit ports are never located directly opposite each other).

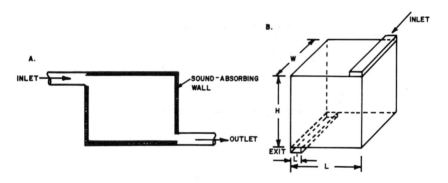

Figure 15.4 A. Single plenum chamber design.
B. Three-dimensional illustration of a single plenum.

Figure 15.4B illustrates the geometry of a single-plenum. The transmission loss associated with a plenum can be approximated by:

$$L_{TL} = -10 \log_{10} \left[A \left(\frac{\cos \theta}{2\pi d^2} + \frac{1-\alpha}{a} \right) \right]^2 \qquad (15.2)$$

where: A = cross-sectional area of exit port (ft^2)
α = random-incidence absorption coefficient of the plenum lining (dimensionless)
a = total lined area in chamber times α (ft^2)
d = distance between inlet and exit port (ft)
$\cos \theta$ = H/d (where H is the height of chamber)
L_{TL} = transmission loss (dB)

Equation 15.2 can be used at high frequencies and small values of L'/L (Figure 15.4). At low frequencies, calculations underestimate actual transmission losses by 5 to 10 dB.

For greater sound reduction, multiple plenum chambers can be employed (Figure 15.5). By knowing the transmission loss for a single chamber (L_{TL_s}) and for a double chamber (L_{TL_d}), the losses for a unit consisting of n chambers can be estimated from:

$$L_{TL_n} = (n-1) L_{TL_d} - (n-2) L_{TL_s} \quad (15.3)$$

where: L_{TL_n} = transmission loss for a plenum with n number of chambers (dB)

n = number of chambers.

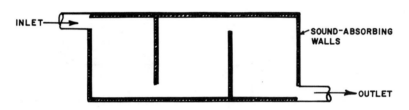

Figure 15.5 Multiple-chamber plenum.

Careful consideration should also be given to the individual components of the system. Proper selection of fans, unit coolers, and mountings should be done in the earliest stages of design. Table 15.2 lists some of the more common approaches to noise control of various units.

Engineering judgment must be used in following established guidelines and in selecting the proper NC curve for a particular design. Factors such as worker's attitudes towards noise levels and economics must be carefully weighed.

Table 15.2 Methods of Noise Reduction for Major Ventilation System Components

Unit	Noise Control Technique
Compressors	Isolation from distribution network; should be supported on resiliently supported inertia blocks
Condensers	Damp surfaces by applying a mastic compound
Cooling towers	Isolation from distribution network; towers should be mounted on heavy-duty ribbed Neoprene pads to reduce vibrations
Evaporators	Evaporation pipe between the capillary exit and the evaporator entrance should be enclosed with a heavy flexible tubing
Fans	Isolation; generally located in a plenum
Piping	Pipe connections should consist of flexible couplings and pipes supported by resilient hangers
Pumps	Isolation from distribution network and mounting on resilient pads

REFERENCES

1. Beranek, Leo (Ed.) *Noise Reduction* (New York: McGraw-Hill Book Co., 1960).
2. Beranek, Leo (Ed.) *Noise and Vibration Control* (New York: McGraw-Hill Book Co., 1971).
3. Blake, M. P. and W. S. Mitchell (Eds.) *Vibrations and Acoustic Measurement Handbook* (Spartan Books, 1972).
4. Harris, C. M. *Handbook of Noise Control* (New York: McGraw-Hill Book Co., 1957).
5. Industrial Acoustics Company. *Duct Silencers*, Bulletin 1.0301.2 Industrial Acoustics Co., Bronx, New York 10462. (1974).
6. Pearsons, K. S. *Handbook of Noise Ratings*, National Technical Information Service, U.S. Department of Commerce, NTIS repl. No. NASA CR-2376 (April 1974).
7. Rettinger, M. *Acoustic Design and Noise Control* (Chemical Publishing Company, 1973).
8. U.S. Department of Housing and Urban Development. *Noise Abatement and Control: Department Policy Implementation Responsibility,*

and Standards, HUD Circular 1390.2 (August 1971).
9. U.S. Environmental Protection Agency. *Information on Levels of Environmental Noise Requisite to Protect Public Health and Welfare with an Adequate Margin of Safety* (Washington, D.C.: Government Printing Office, 1974).
10. Wells, J. R. "Acoustical Plenum Chambers," *Noise Control* 4:9 (1958).

CHAPTER 16

INSTRUMENTATION FOR NOISE ANALYSIS*

Noise measurements usually fall into one of two categories. The first category consists of sound level measurements usually made of the dBA level, e.g., OSHA surveys and product noise ratings. Common measuring instruments are sound level meters and dosimeters. The second category consists of analyses made in support of environmental noise reduction programs and product design programs. In addition to measurement of dBA level, these analyses require frequency analysis or amplitude distribution analysis.

MICROPHONES

Microphones are an essential part of all acoustical measurements, and microphone characteristics control the sections of sound level meter specifications and ANSI standards that deal with such parameters as frequency range, directivity, temperature stability and effects of humidity. In fact, the microphone is the main distinguishing factor among different types of sound level meters.

Ideally, the microphone should produce an electrical signal that is an exact replica of the acoustical disturbance. It must operate over a wide dynamic range and a wide frequency range, and it must be stable under severe changes in environmental conditions. Of the many different principles of microphone construction, only condenser and piezoelectric microphones are used for instrumentation purposes. The condenser microphone can be of the stretched metal diaphragm or the electret type. The stretched metal diaphragm condenser microphone is the best choice

---

*By Anthony J. Schneider, B & K Instruments, Inc., Cleveland, Ohio, and David Marsh, Bruel & Kjaer, Copenhagen, Denmark.

for accurate and repeatable measurements, while the piezoelectric and electret condenser are second-best choices if price is important.

The construction of a stretched metal diaphragm microphone, commonly called a condenser microphone, is shown in Figure 16.1. The microphone cartridge consists of a thin metallic diaphragm in close proximity to a rigid backplate. These two elements are electrically insulated from each other and form the plates of a capacitor. A DC polarizing voltage is applied across plates of the plates. Variations in pressure due to sound waves will move the diaphragm, thus varying the width of the air gap.

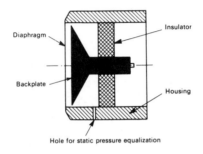

A

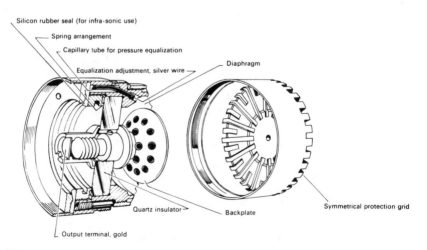

B

Figure 16.1 A. Schematic construction of the condenser microphone.
B. Construction of a condenser microphone.

Consequently, an alternating charge is generated on the capacitor. By careful design it is possible to keep the electrical output proportional to sound pressure over a wide range of frequencies and dBA levels. In order that changes in atmospheric pressure will not change the static position of the diaphragm and cause a change in microphone sensitivity, the air inside the microphone is vented to atmosphere. Therefore, the static pressure remains the same on both sides of the diaphragm. A common temperature coefficient for the metals in the diaphragm and housing is selected so that temperature stability of the microphone can be guaranteed for long periods.

There is another type of condenser microphone, which is commonly called the "Electret" microphone. Its schematic construction, shown in Figure 16.2, is similar to a stretched metal diaphragm condenser microphone in that it has a diaphragm separated from the backplate by a thin air gap. The diaphragm of an electret microphone, however, is made up of a polymer film whose outside surface is metal plated. The plating and the backplate form the electrodes of a condenser with each raised section of the metal plate electrode representing a miniature microphone. The outputs are all summed to produce the effective output of the microphone. Since the polymer has electrical charges embedded into it, thereby containing its own charge, no polarization voltage is required.

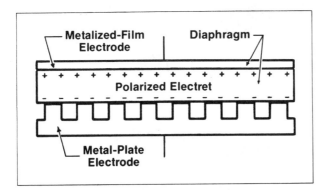

Figure 16.2 Schematic diagram of an Electret microphone.

The most common form of a piezoelectric microphone uses a piezoelectric ceramic element to which a bending movement is applied when the diaphragm is exposed to sound pressure. An example of this type of microphone is shown in Figure 16.3. The ceramic bender element is supported at both ends. The conical diaphragm applies a force to the center

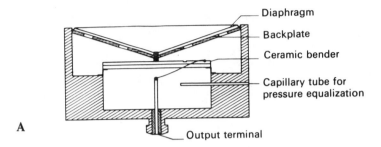

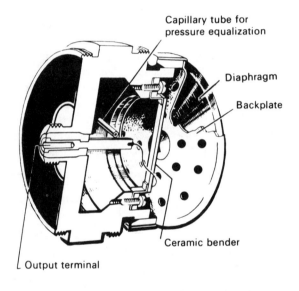

Figure 16.3 **A.** Schematic drawing of a piezoelectric microphone.
B. Practical construction of a piezoelectric microphone.

of the bender, thus producing an output voltage proportional to the amplitude of the diaphragm motion.

Piezoelectric microphones are often used in general purpose sound level meters. They are relatively inexpensive and have the advantage that they do not need a power supply. For precision sound level meters, however, piezoelectric microphones are generally not suitable. Size-for-size, piezoelectric microphones have poorer frequency response and lower sensitivity than condenser microphones.

MICROPHONES AND THE ENVIRONMENT

An instrumentation microphone should operate in a sound field without affecting the progress of the sound wave. One definition describes sound pressure as the pressure that would exist at a point in space if a microphone were not present. Unfortunately, the 1-inch and $1/2$-inch microphones commonly used on sound level meters do affect progress of the sound wave. The interference occurs at higher frequencies where the short acoustical wavelengths approach the diameter of the microphone.

There is a large additive effect when the sound wave is perpendicular to the plane of the microphone's sensing surface. For other angles of incidence, the sensitivity is lower. As a result, a microphone is omnidirectional at low and mid-frequencies, but it develops a large directional effect at higher frequencies. Figure 16.4 shows a typical family of corrections to the pressure-response curves of so-called random incidence, or grazing incidence, microphones. (A pressure field is one in which there is no wave propagation.) The name of this class of microphone is derived from its flat frequency response in the random sound field of a reverberant room and in a free field where the microphone is oriented so that the sound wave grazes across the face of the microphone.

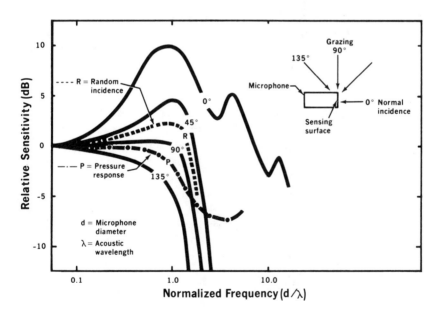

Figure 16.4 Effect of microphone interference on the sound field.

However, in factory and community environments where more than a single sound source is present, sound waves more perpendicular to the diaphragm may be heavily amplified and those coming from behind the microphone may be attenuated. Maximum amplification occurs when microphone diameter equals the wavelength of sound. For a 1-inch microphone this occurs at about 13.6 kHz. Because the pressure response begins to roll off at a somewhat lower frequency, the actual directional response curves peak out at 5 to 8 kHz, depending upon the design of the individual microphone. So the potential for overestimating sound level with this class of microphone lies in the range of 1 to 8 kHz. Unfortunately, this is the frequency range emphasized by A-weighting filters used in the measurement of dBA.

An alternate microphone is the so-called free-field or perpendicular-incidence microphone. It is designed to have an over-damped pressure-response curve in order to equalize the perpendicular incidence error discussed above. This class of microphone has a flat frequency response for sound waves that are perpendicular to the plane of the sensing surface. Hence, its advertised axis of response is also its most sensitive axis. It has the advantage that it can never overestimate sound level, regardless of angle of incidence. In fact, the microphone can be oriented for maximum reading in order to get the most accurate measurement of sound level. A partial family of directivity curves for a grazing ($90°$) incidence and a perpendicular ($0°$) incidence microphone are shown in Figure 16.5. In many designs of microphones, the directional effects are even greater.

Because OSHA environments consist of many intermittent sound sources, maximum accuracy is achieved if the microphone is truly omnidirectional over the entire frequency range of interest. This can be achieved by using a tiny microphone, but the reduced sensitivity of smaller microphones makes this solution impractical. An alternative is to use adaptors that eliminate directional characteristics by making the microphone appear to be smaller than it actually is. Two such combinations of microphone and adaptor are shown in Figure 16.6. In each case, the limits of response frequency for any angle of incidence are held to within 2 dB.

Microphones are often used with wind screens. Although designed to reduce the effect of wind noise for measurements outdoors, they also serve to protect the microphone from oil spray, flying chips and dust during factory noise measurements.

Good quality microphones will be stable over the period of measurement, but the data should be backed up by an acoustical calibration before and after each day's use. Humidity is seldom a problem, with the provision that moisture must not condense on the sensing surface.

INSTRUMENTATION FOR NOISE ANALYSIS 311

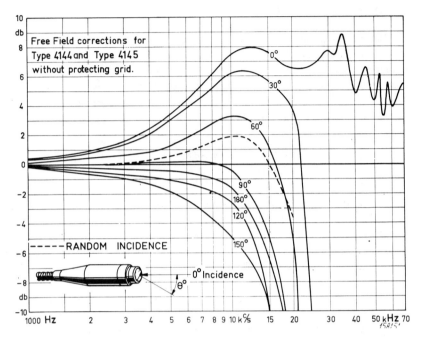

Figure 16.5 Free-field corrections for microphone without protecting grid (electrostatic actuator method of pressure calibration).

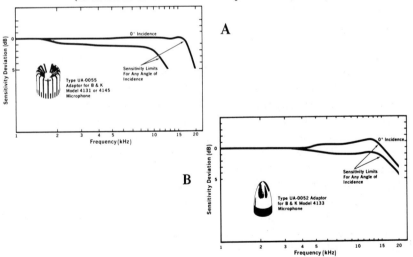

Figure 16.6 A. Special adaptor gives good omnidirectional characteristics for 1-inch condenser microphone. B. Nose cone adaptor gives good omnidirectional response for $1/2$-inch condenser microphone.

SOUND LEVEL METERS

A sound level meter is an acoustical measuring instrument consisting of a microphone, amplifier, weighting filters and readout meter, as shown in Figure 16.7. It will meet one or more governing standards: ANSI S1.4-1971 Specification for Sound Level Meters dominates in the United States, but IEC-179, an international specification for a precision sound level meter, is often specified for more exacting work. S1.4-1971 specifies three types of sound level meters:

Type 1	precision sound level meter
Type 2	general purpose sound level meter
Type 3	survey sound level meter

The standards outline straightforward electrical specifications such as meter damping, response of frequency weighting networks and certain environmental characteristics. But the distinguishing factor among the three types of sound level meters is the tolerance allowed on frequency response and directional response. Therefore, the microphone determines whether a sound level meter meets Type 1, 2 or 3 requirements.

The A-weighted frequency-response tolerances of current standards are summarized in Figure 16.8 for the important range about 1 kHz. Note that precision sound level meters can have no response above 15 kHz, and general purpose and survey sound level meters are guaranteed to operate only up to 10 kHz. Most users demand more than the guaranteed minimum, so the prospective user must check the specifications for individual sound meters.

Of much greater importance to the user are the directional-tolerances shown in Figure 16.9. These curves are normalized so that conformance to the A-filter by main-axis measurements under free field conditions is represented by a flat response. The ANSI standards permit use of grazing incidence microphones, which overestimate sound waves with perpendicular incidence to the sensing surface. The Type 3 sound level meter is permitted to have positive errors ranging up to 14 dB at 8 kHz. To protect the user, OSHA has disapproved use of Type 3 meters for operator exposure surveys. The tolerances for Type 2 meters are much tighter. A typical Type 2 meter for general purpose work is shown in Figure 16.10.

Tolerances for Type 1 meters are similar to Type 2, but are extended to 15 kHz. Many organizations are not satisfied with Type 1 directional tolerances so the IEC-179 standard is often specified. The IEC-179 standard essentially requires that the most sensitive axis coincide with the main axis response. The intent is to specify a perpendicular-incidence microphone that can never overestimate sound level.

INSTRUMENTATION FOR NOISE ANALYSIS 313

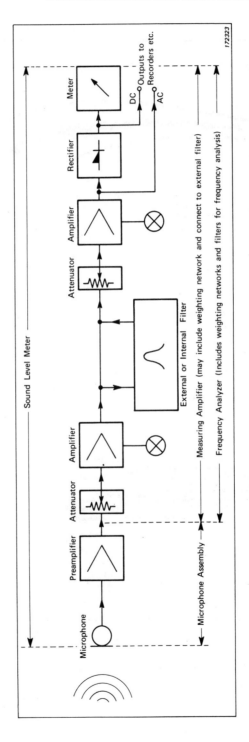

Figure 16.7 Block diagram of the important elements of a sound level meter.

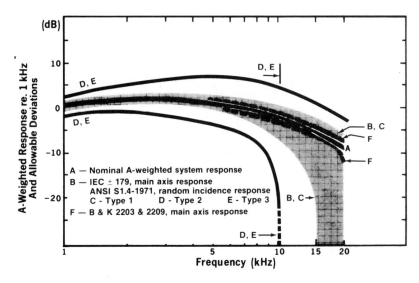

Figure 16.8 Current standards permit wide tolerances in A-weighted frequency response of sound level meters. However, close adherence of an instrument to the "A" curve provides protection against severe amplification or attenuation of the data obtained in using the instrument in noise-measurement programs.

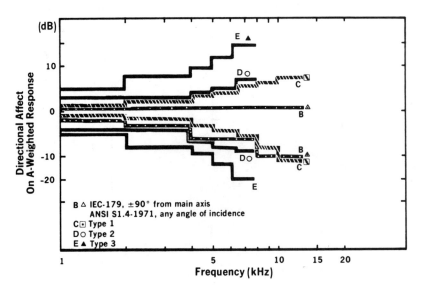

Figure 16.9 The standards permit irregularities in the directional characteristics of sound level meters. Industrial noise measurements generally require microphones that are more omnidirectional than some standards guarantee.

Figure 16.10 A typical Type 2 sound level meter for general-purpose noise investigations. The instrument shown here is light in weight and can be operated in one hand.

One other distinction in sound meters should be made. If the meter is to be used for noise reduction programs, a meter should be selected that has optional octave filters available. These meters are generally more expensive than those designed solely for noise surveys.

Sound level meters measure the rms level of the acoustical signal. Two meter damping positions are required—"fast" and "slow"—corresponding to averaging times of approximately 200 and 800 ms. Even in "slow" response the meter will often fluctuate. In this case the operator must note the range of levels indicated on the meter. For transient noise he notes the maximum reading during the event. But OSHA imposes a measurement requirement for impact noise that is not covered in the ANSI and IEC standards. For impact noise such as forging, stamping and press operations, an OSHA noise limit is imposed based upon the peak sound pressure level. A peak-responding detector with a rise time not exceeding 50 microseconds is needed to measure this, and measurements must be made in the C or linear response mode. Furthermore, the meter must have a hold-circuit that captures the peak level and decays at a rate less than 0.05 dB/sec. This capability may be built into the sound meter or the manufacturer may supply a separate accessory. A precision sound level meter with peak-hold feature and optional octave filter is shown in Figure 16.11.

316 INDUSTRIAL NOISE CONTROL HANDBOOK

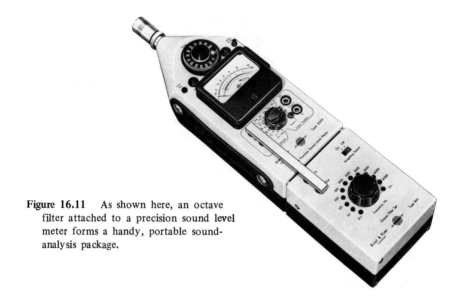

Figure 16.11 As shown here, an octave filter attached to a precision sound level meter forms a handy, portable sound-analysis package.

CALIBRATION

Common practice and most industry standards require that sound level meters be calibrated with an acoustical signal before and after each day's use. Most calibrators use an electrical signal to drive a diaphragm that serves as a loudspeaker in the calibration cavity. Because the calibration level is a function of the applied voltage, a regulating circuit is used to maintain the supply voltage at a constant level. Most calibrators of this type operate at 1 kHz. At this frequency the weighting filters have no gain, offering the advantage that a 1 kHz calibrator can be used with the sound meter in the A-weighting mode without using any correction factors. A typical electrically controlled acoustical calibrator is shown in Figure 16.12. Here a piezoelectric ceramic element is used to drive the metallic diaphragm, which acts as a loudspeaker.

THE ACOUSTICAL MEASUREMENT

Today's acoustical measurements are made either to evaluate an environment or to measure the noise output of a machine. Generally, there are several noise sources in the area, and the operator's machine may not be the major noise source. It is important that the microphone for measuring operator environment be omnidirectional—that it accurately measures all sound sources, regardless of their direction from the operator.

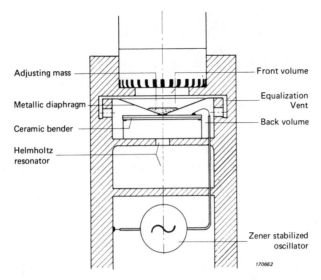

Figure 16.12 Cut-away view showing principle of operation of a portable acoustical calibrator.

In contrast to operator position measurements, machine measurements call for measuring only the noise emanating from a single source. Machine measurements must usually be made on the factory floor, particularly frequency analyses in support of enclosure design. In these cases, measurements must be made close to the machine to ensure that the machine sound dominates the environment. This measurement is best made when all other machines are shut down, such as during nonworking periods, so that the most accurate characterization of the machine can be made. With the exception of dose measurements for which mobile employees wear a microphone on their lapel, the person making the measurement should be behind the microphone so that his own body does not act as a reflecting object to disrupt the measurement.

DOSIMETERS

The OSHA law requires characterization of varying noise environments by a single number criteria. The environment may be that experienced by a mobile employee whose work takes him into many different factory areas, or it may be a fixed work area where intermittent machine operations produce unpredictable variations in noise level. OSHA specifies a maximum daily noise dose (D) of unity:

$$D = \frac{C_1}{T_1} + \frac{C_2}{T_2} + \ldots + \frac{C_n}{T_n}$$

where C is the total exposure at a given steady dBA level and T is the maximum allowable exposure time at that level during a 8-hour work day. The relationship between dBA level and allowable duration are shown in Figure 16.13.

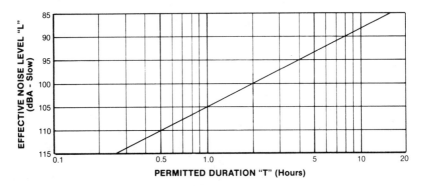

Figure 16.13 OSHA (noise) table of dBA versus allowable exposure for working personnel.

Noise dose can be predicted for mobile workers by studying their movements and estimating the total time they are exposed to each different dBA level they experience. Alternatively, dosimeters can be worn by mobile employees and by the stationary employees to obtain a direct measure of noise dose. Dosimeters are integrating sound meters that operate over fixed time periods, usually 8 hours. They convert dBA level to frequency in accordance with the OSHA dBA/exposure time relationship and produce a composite exposure number on a digital readout. Their advantage is that they are simple instruments that convert time-varying sound level to an equivalent steady level. A typical OSHA dosimeter is shown in Figure 16.14a and b, and a block diagram is given in Figure 16.15.

FREQUENCY ANALYZERS

Basic sound level meters are commonly used to determine if there is an OSHA violation of steady noise exposure or if the overall noise level of a machine meets its guarantee. As such, they are go-no go monitors. The readings give little assistance to the person responsible for reducing

INSTRUMENTATION FOR NOISE ANALYSIS 319

Figure 16.14a A typical OSHA noise dosimeter. It can be worn conveniently by the worker in his shirt or coat pocket or attached to his belt.

Figure 16.14b As shown here, an OSHA dosimeter can be worn in a worker's pocket with an extension microphone located near his ear to pick up the effective noise of his working environment.

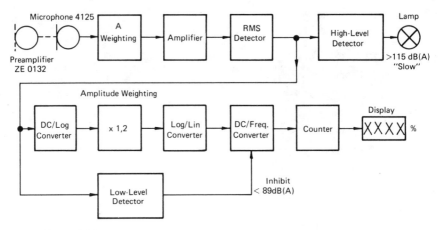

Figure 16.15 Simplified block diagram of a personal noise dose meter, complying with OSHA criteria.

noise levels. He must usually know the frequency content of the sound because the design of efficient barriers and enclosures is heavily dependent upon the frequency spectrum and absorbing materials on walls or ceilings only work if high frequencies dominate.

Because sound level meters are portable instruments, portable filters are also required. Filters take the form of an octave or a $1/3$-octave (see Figure 16.16) sequence of bandwidths that permit the wide audio

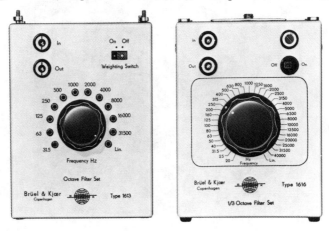

Figure 16.16 An octave filter set (left) or a third-octave filter set (right) can be attached quickly to a precision sound level meter to provide a versatile, portable sound-analysis package. (See Figure 16.11 for a combined system.)

range to be reduced to a practical number of segments. The center frequencies and bandpass characteristics of octave and $1/3$-octave filters are controlled by ANSI S1.11-1966. A set of octave response curves are shown in Figure 16.17. Although analyzers used in the laboratory contain a large number of individual filters operating in parallel, portable analyzers consist of a single filter that the operator tunes or steps from one position to the next. A typical octave filter, attached to its sound level meter, is shown in Figure 16.11.

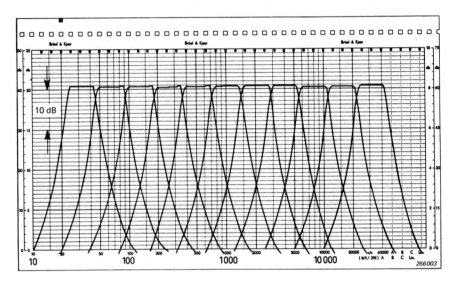

Figure 16.17 Graphic recording of the frequency characteristics of an octave filter set.

Historically, sound level meters have been designed so that octave analyses can be made only when the meter is in its flat frequency-response mode, thus preventing analysis of A-weighted signals. But today almost all noise reduction tasks are directed toward reducing dBA level. Therefore, it is necessary that today's instruments permit series-connection of the A-weighting filter and the octave filters. In this way, the dominant A-weighted octave band is directly identified. A typical octave analysis is shown in Figure 16.18. The dBA level is plotted in octave bands to show the beneficial effects of lining the enclosure with sound-absorbing materials.

322 INDUSTRIAL NOISE CONTROL HANDBOOK

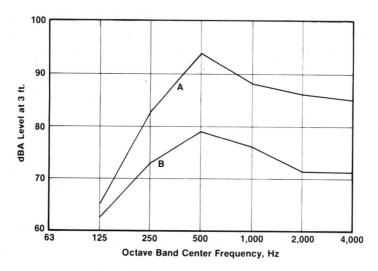

Figure 16.18 Noise level 3 feet from an enclosed electric motor. Design A is a standard enclosure; design B is lined with sound-absorbing material.

AMPLITUDE DISTRIBUTION ANALYZERS

As octave analyzers supplement sound level meters in programs to reduce dBA level, the dosimeter also needs a support instrument to identify dominant machines. Management needs an economical solution to noise problems; therefore, dominant noise components must be identified before any noise reduction program is initiated. The machine that is the noisiest one in the shop when it operates is not necessarily the dominant noise source in a mixed noise environment. Its contribution can only be determined by integrating the total time that the machine operates during the day. And, it is inappropriate to build an enclosure for a noisy machine if an equal noise reduction can be achieved by reducing background noise, such as by conveyors or metal tote pans.

Figure 16.19 shows a typical instrument that can be used to identify dominant noise sources. It classifies dBA levels into 12 dBA bands over the OSHA noise range and totalizes the accumulated noise level in each amplitude band.

One application of the Environmental Noise Classifier is in the measurement of the OSHA mixed exposure level where it serves as an area dosimeter *and* as a guide to the most economical solution to noise reduction problems. Consider the data shown in Figure 16.20. The coefficient values are the ratios of the time spent in each band divided by the

INSTRUMENTATION FOR NOISE ANALYSIS 323

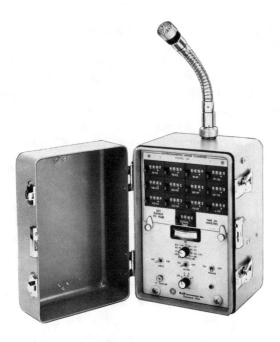

Figure 16.19 An Environmental Noise Classifier serves to monitor noise and classifies the dBA levels into 12 dBA bands over the OSHA noise range and gives the total accumulated noise level in each amplitude band.

dBA LEVEL	MAXIMUM HRS/DAY	CASE 1		CASE 2	
		ACTUAL HOURS	OSHA COEFFICIENT	ACTUAL HOURS	OSHA COEFFICIENT
< 90	–	2.5	–	–	–
90 – 92	6	3	0.5	1	0.17
92 – 95	4	2	0.5	1	0.25
95 – 97	3	0	0	2.4	0.80
97 – 100	2	0.5	0.25	0	0
OSHA MIXED EXPOSURE LEVEL			1.25		1.22

Figure 16.20 Example of amplitude analysis of OSHA environments.

maximum allowable time in the band during an 8-hour day. If the sum of all coefficients (mixed exposure level) exceeds unity, the noise is in violation of the OSHA code. In each example, the times spent in the individual bands are read from the Environmental Noise Classifier in minutes and recorded in the table in hours. In each example, the mixed exposure level is about 1.25.

Without going into great detail, it can be seen that quieting the obvious machine that operates for $1/2$-hour at the 97-100 dBA level in Case 1 is not an economical solution to the problem. If reduced to below 90 dBA, the OSHA coefficient would still be 1.00. Further, quieting a large machine by 7 to 10 dBA is a major undertaking and may require construction of costly enclosures. The enclosures create problems of their own, such as access, increased labor and temperature build-up. The data suggest the reduction of ambient noise and lower level machine noise to accomplish the goal.

In Case 2, there is no choice but to work on the noisiest machine. But the histogram suggests how many dB the level must be reduced. In order to reduce the mixed exposure level to less than unit, the offending machine must be reduced to the 90-92 dBA band, or a reduction of 3 to 5 dBA. The same reasoning can be applied to reducing noise of machine tools where the sound level changes with each cycle.

CHAPTER 17

AUDIOMETRIC TESTING AND DOSIMETERS

AUDIOMETERS

The periodic testing of the hearing ability of persons is called audiometry. In industry, such testing may be carried out by a physician, qualified technical personnel, or anyone who will use techniques that meet the minimum requirements of OSHA. The major instrument used in audiometric testing is the audiometer, which measures a person's hearing ability. To comply with **OSHA**, audiometers must meet the specifications for limited range, pure tone audiometers as set forth in ANSI Standard S3.6-1969, which is titled "Specifications for Audiometers."

When measuring the threshold hearing levels of industrial personnel, the limited range audiometer is most suitable. Under the ANSI specifications, limited range audiometers must be capable of producing tones of the following frequencies: 500, 1000, 2000, 3000, 4000 and 6000 Hz from 10 dB to 70 dB. This must be done for air conduction tests, and bone conduction may be neglected.

The ANSI Standard does not give preference to either automatic or manual audiometers; it permits the use of either. An operator is required to be present throughout the audiometric test when a manual audiometer is being used because he must manually present the various intensities at specific frequencies. The operator will establish the threshold levels at each frequency and will manually record them by means of the subject's response or lack of response.

When automatic audiometers are used, the test subject will respond to the intensity of the tones given to him by releasing or pressing an electric switch. When there is no response, the audiometer will increase the intensity of the signal until there is a response, and then the intensity will decrease. This process automatically records the subject's threshold levels as they are obtained. With the automatic audiometer an operator is only needed to explain the test to the subject, position the earphones, and press the starting button. Table 17.1 gives the important aspects of both types of audiometers.

Table 17.1 Distinguishing Characteristics of Manual and Automatic Audiometers

Factors Considered	Relative Merits	
	Manual Audiometer	Automatic Audiometer
Time of test	Depends somewhat upon operator skill.	Usually 30 sec per frequency for each ear automatically controlled (6-7 minutes per subject).
Operator's time required	Operator must explain test, fit earphones, present tones at various levels of intensity at each frequency, and record thresholds. Trial period may be given before test to be sure subject understands.	After explanation and fitting of earphones, operator pushes start button. Instrument automatically presents tones, varies intensity and frequency and records thresholds. Trial period may also be given.
Threshold judgments	Made by operator based on subject response.	Recorded by instrument based on subject response.
Simultaneous testing of more than one person	Only one person can be tested at a time.	Limited only by number of audiometers and/or testing booths; however, four or five subject limit is practical.
Retest	Operator can retest a particular frequency immediately should malingering or inattentiveness be suspect.	Retest can be given at a particular frequency after complete test, or by overriding the automatic if operator has test under surveillance.
Other	Results affected by operator fatigue and possible recording errors.	Instrument is not biased. Uniformity of tests. Test-retest reliability; less operator training required.

MAINTENANCE OF ACCURACY

Before being used for testing purposes, an audiometer must be checked to be accurate in such qualities as frequencies, purity of tones and sound pressure levels. A standard audiometer should be calibrated at least once every year by an expert, and, in accordance with OSHA, it must be biologically checked at least once a month, or before each use if used less than once a month. This is accomplished by having a subject with a known hearing ability take an actual test, the results of which are then compared to previous tests for stability. The company's personnel can perform this test and the results should be preserved for possible inspections.

The audiometer room or booth in which the subject is placed during testing (see Figures 17.1 and 17.2) should be inspected regularly for damaged gaskets and seals. Also, sound level readings should be made occasionally to determine if the sound attenuation qualities of the structure are up to par.

Figure 17.1 Typical audiometric room.

Figure 17.2 Industrial audiometric booth.

AUDIOMETRIC TESTING OUTSIDE THE PLANT

In cases where audiometric tests are conducted outside the plant, the Department of Labor representative will inspect the facilities and test records. The plant management will make arrangements for such inspections with the person conducting the audiometric tests and may be present with the representative throughout his inspection and examination of the records.

THE RECORDS

There must be a statement on the audiometer that tells whether it is calibrated to ASA 1951 or to the current American National Standard. A log of all accuracy and biological checks of the audiometer must be kept available for inspection. Also, a record of each audiogram made on each worker must be filed and maintained for inspection. The records of all the audiometric tests will show whether readings are based on ASA 1951 or on the American National Standard, which is identical to the ISO 1964 Standard. The records of every worker required to be tested must be kept for one year after termination of employment, or after the worker has been transferred to an area with noise levels below 90 dBA.

All records shall be inspected for evidence of any deterioration of hearing ability and of the action taken to prevent further deterioration in the employees who have suffered some hearing impairment.

AUDIOMETRIC TEST BOOTHS

The site for the audiometric testing is important as the test tones can be masked easily by ambient noise. Unless an extremely quiet room is available, away from traffic, machines, people, conversations, and doors, an audiometric testing booth is required to conduct an accurate test.

In accordance with OSHA, the ambient noise level for conducting tests must meet the criteria as set forth in ANSI Standard S3.1-1960 (with latest revision) "Standard for Background Noise in Audiometric Rooms." This standard lists the maximum allowable sound pressure levels of the ambient noise at each frequency of testing. Table 17.2 presents the allowable levels as given in the referenced standard.

Each location must first be checked for "quietness" with a noise level meter utilizing either $1/3$-, $1/2$-, or full-octave band settings. For example, if the $1/2$-octave band setting of a sound level meter is used for the range setting including 1500 Hz, the allowable maximum ambient level cannot

Table 17.2 Maximum Allowable Sound Pressure Levels (dB Ref. 0.0002 microbar) for No Masking Above Zero Hearing Loss Setting of Audiometers[2]

Test Frequency of Audiometer (Hz)	Octave Band	$1/2$-Octave Band	$1/3$-Octave Band
125	40	37	35
250	40	37	35
500	40	37	35
750	40	37	35
1000	40	37	35
1500	42	39	37
2000	47	44	42
3000	52	49	47
4000	57	54	52
6000	62	59	57
8000	67	64	62

exceed 39 dB for audiometric testing. A sound level meter must be used with a frequency analyzer having $1/3$-, $1/2$-, or full-octave band settings. These can be borrowed or rented, and in some cases vendors of audiometric equipment will conduct a site survey at no charge. If any table value(s) is exceeded, the site is not good enough for audiometric testing. Therefore, either the noise level must be reduced, an audiometric booth utilized, or another site selected. The test must be made when all machinery present is operating and noise level is at its highest.

One must remember that audiometric rooms or booths do not eliminate noise, they only attenuate it. The more attenuation required, the more material required for fabrication, and hence, the greater the cost. Audiometric rooms are available in all sizes, from smaller than telephone booths to 6 x 10 ft rooms and larger for multiple person testing. Some booths are mounted on wheels or casters, while most are a permanent type that is too big to be moved through doorways. However, assembly can be performed by plant maintenance men, who must be sufficiently supervised, or suppliers of such units can provide installation crews at some cost.

As soon as size of an audiometric room is established, the adequacy of its attenuation qualities can be determined by requesting attenuation data from the vendor before purchase. This data should contain the attenuation in dB of the booth or room at each frequency at which testing will be performed. A simple calculation can then be made for each test frequency as follows:

$$\frac{\text{Ambient noise level (in dB)} + 10 \text{ dB}}{\text{Safety factor-booth attenuation}} = \frac{\text{Sound level inside}}{\text{audiometric room}}$$

This calculation can be repeated for each frequency.

A 10-dB safety factor is usually recommended to ensure a margin for any later increase in the ambient noise. The values calculated for sound pressure level at each frequency should then be compared with the maximum allowable values shown in Table 17.2. If no values are exceeded, the audiometric room is satisfactory for test purposes. If one or more values are exceeded, then the room is unsatisfactory at that specific site, and either the site must be changed, the noise abated, or another room selected with better attenuation qualities.

Rooms or booths can come with many options; standard features are usually one window, a light, and connections for the audiometer in addition to a door, walls, ceiling and floor. Costs vary directly in proportion to attenuation and size. Extras range from an outside shelf for the audiometer to fluorescent lighting to rugs on the floor. However, two options that should be considered are vibration isolator rails to attenuate vibration transmitted through the ground and a blower-ventilation system.

ALTERNATE PROGRAMS

An alternative to setting up one's own audiometric program is to design a program with either a hospital or clinic or a mobile laboratory company. The cost of any hospital or clinic arrangement usually depends upon the number of persons requiring testing. Although fees could run as high as $15 per person for small groups, this may be the best method for companies with a small number of employees to test. However, employee time lost must be considered because the employees must leave the plant, travel to and from the hospital or clinic, and perhaps even wait to be tested.

Several companies offer in-plant testing of employees by means of mobile vans equipped with audiometers and booths. The vans come to the plant site, take sound level measurements to certify the site and then give the employees the test, often at a rate of 20 subjects per hour, utilizing multiple instruments. Prices could range from $10 per person tested for small groups to $6 per person for several hundred employees tested. Some companies require a minimum of 25-50 people to be tested before they will send out their van.

The mobile van companies will do all the work, qualify the site, provide the technicians, maintain records, submit reports, and analyze the test results. All the company has to do is to schedule the people to be tested. However, certain problems often arise. For example, some people might be absent the days the van is on-site. Also, there is the problem of testing shift workers, and those people joining the company after the van has left. All those missing the test can be referred to a hospital or clinic. Added to this is the problem of "minimum time away from noise" prior to taking an audiometric test. A hearing test taken after exposure to noise can result in a temporary or apparent threshold shift and may appear to be a hearing loss or a permanent threshold shift. Authorities differ on the precise amount of rest time from noise before testing, but most agree on at least 16 hours. If followed, this presents a serious scheduling problem, particularly if utilizing a mobile van. And the worker should be kept away from noise that day until the time of his test.

If a company has its own equipment, it can schedule a few tests each morning before the person has entered the noisy area for work. This is even more feasible if the company will be utilizing existing employees to supervise the testing (such as nurses, medical technicians, personnel department people) as testing will only take up a small portion of their day. Such a program distributes the testing load more evenly throughout the working year.

DOSIMETERS

Dosimeters are portable instruments that can be attached to a worker and go with him throughout his normal working tasks. The dosimeter measures and records the full noise exposure of the employee and at the end of the work shift gives the percentage of allowable noise exposure received by him. In effect, it computes the dosage/time equation for the entire working day. This equation is as follows:

$$\frac{C_1}{T_1} + \frac{C_2}{T_2} + \frac{C_3}{T_3} + \ldots \frac{C_n}{T_n}$$

where C_n is the total time of exposure at noise level n, and T_n is the total time permitted at that particular noise level. If the sum of all of these fractions is greater than one, then the combination of exposures has exceeded the permissible limit.[3]

Dosimeters are slightly larger than a pack of cigarettes and weigh about 8 oz. Powered by two small batteries, they are completely portable and can be easily clipped to a shirt pocket or a belt, as shown in Figure 17.3.

Dosimeters contain a microphone for sound pickup, and some models use a built-in microphone. Various models have a microphone attached

Figure 17.3 A portable dosimeter in worker's shirt pocket.[4]

to the dosimeter through a lead-in wire, which makes it possible for the microphone to be placed anywhere on the employee. Therefore the microphone can be placed in the ear like a hearing aid as shown in Figures 17.4 and 17.5.[3]

There are many types of dosimeters offered on the market. In choosing the best type for one's purposes, several factors should be taken into consideration. Individual variations should be carefully examined in the light of a company's overall OSHA compliance program, conservation program and individual personnel policies. Another factor to consider is whether a dosimeter with an integral readout is needed. Dosimeters with integral readouts continuously integrate and display the percentage of daily noise exposure. These are self-contained and do not require the aid of any additional readout equipment. In the dosimeter illustrated in Figure 17.6, the exposure data can be read anytime, without erasing any stored data. On this model the separate noise-exposure

AUDIOMETRIC TESTING AND DOSIMETERS 333

Figure 17.4 Subject wearing a dosimeter with the microphone inserted in his ear like a hearing aid.[4]

Figure 17.5 The microphone of the portable dosimeter fits into the ear.[4]

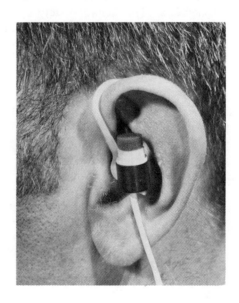

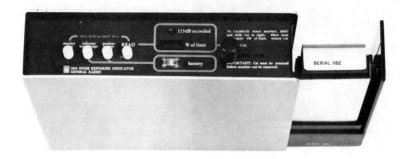

Figure 17.6 Integral readout dosimeter.[4]

indicator has a built-in acoustic calibrator that permits checking of its monitor calibration before and after measurements are made. The monitor slides onto the indicator, and one indicator can service any number of monitors. The monitor-battery condition and the displays on the indicator can thus be checked. In addition, the indicator is battery-operated so that it can be carried to any work area to check exposure conditions.

An important factor in choosing dosimeters is economics. If only one or a few dosimeters are needed, the single unit type is more economical. If many dosimeters are to be used continuously at one locale, the two-instrument system can be less costly since only one readout indicator will handle any number of measurement units. However, this cost advantage must be weighed against the fact that operation is not as convenient as with the single-unit system. Also, additional maintenance is usually necessary to prevent the readout connections from becoming dirty or corroded, causing erroneous readings.[5]

Another factor in comparing dosimeter costs is how the unit is powered. Some units come with a rechargable nickel-cadmium battery that will last at least a year, providing a minimum of 8 hours use per charge. However, other dosimeters use disposable batteries that require frequent replacement.

REFERENCES

1. I.D.E. Processes Corporation, Kew Gardens, New York 11415.
2. American National Standards Institute, ANSI 53.1-1960 (R-1971).
3. Cheremisinoff, P. N. and R. P. Angellilo. "Noise Dosimeters," *Pollution Eng.* 6(6):32-34 (1974).
4. General Radio Corporation, West Concord, Massachusetts.
5. Edmont-Wilson Division of Becton, Dickinson & Company, Coshocton, Ohio.

CHAPTER 18

NOISE LEVEL INTERPOLATION AND MAPPING*

INTRODUCTION

Industrial noise is a common cause of hearing loss among workers. In addition, industry-generated noise that exceeds community ambient noise levels is a significant annoyance and may evoke community reaction.

NOISE LEVEL AND FREQUENCY WEIGHTING

Sound is characterized by its frequency and pressure level. Sound pressure level L_p (decibels) is given by the logarithmic relationship

$$L_p = 10 \log_{10}(p^2_{rms}/p^2_{ref}) \tag{18.1}$$

where p^2_{rms} is the mean square sound pressure (Newtons/m^2) and p_{ref} = 2×10^{-5}. Alternately, intensity level L_I (decibels) is given by the logarithmic relationship

$$L_I = 10 \log_{10}(I/I_{ref}) \tag{18.2}$$

where I is the sound intensity in watts/m^2 and $I_{ref} = 10^{-12}$. For airborne sound and ordinary conditions, the difference between L_p and L_I is negligible.

Audible sound covers the frequency range from about 20 Hz to 20,000 Hz and ranges in sound pressure level from about L_p = 0 dB (the threshold of hearing) to L_p = 140 dB or higher (the threshold of pain). The A-weighting network in a sound level meter combines the effect of sound at various frequencies approximately as the human ear responds. The A-weighted sound pressure level in decibels (dBA) is a convenient and widely used measure of industrial noise. For purposes

---

*By Charles E. Wilson, Mechanical Engineering Department, New Jersey Institute of Technology, Newark, New Jersey.

of hearing conservation and assessment of community noise, measurements in dBA are generally adequate except when pure tones are prominent.

INDUSTRIAL HEARING DAMAGE RISK

Routine hearing level tests are made with an audiometer, with hearing thresholds measured at several frequencies. Criteria for acceptable noise exposure depend on the definition of a hearing handicap. If a hearing handicap is defined as an average hearing loss of 25 dB, continuous (8 hour per day) industrial noise exposure to levels up to 80 dBA will not increase the risk of developing a handicap.[1] This exposure level is based on the results of audiometry examinations of industrial workers and comparisons (by age group) with others who were not exposed to high industrial noise levels. Other research[2] suggests a much lower safe exposure level on the basis of a different hearing impairment criterion.

COMMUNITY NOISE FROM INDUSTRIAL SOURCES

When industrial and residential zones abut, many problems can develop, the most pervasive often being noise. Residents may complain of sleep interference, speech interference, or simply annoyance. At a distance of four feet between talker and listener, speech communication is barely possible at normal male voice level with a background noise level of 63 dBA.[3]

Sleep interference is more difficult to express quantitatively. In one study,[4] it was found that the time required to fall asleep increased by 1 to 1.5 hours at noise levels between 50 and 60 dBA. Furthermore, at those noise levels, the depth of sleep was decreased. Annoyance is a very subjective effect. However, numerous studies have indicated that when intrusive noise exceeds background noise levels by 5 to 10 dBA, community action is likely.

NOISE CONTROL

Principal methods of industrial noise control include reduction of noise output at the source and control of noise transmission paths. Noise sources should be considered first. After reducing noise source output to the lowest practical levels, noise transmission paths through solids and air should be examined. For airborne noise outdoors, buildings and barriers between the noise source and receiver have a substantial effect. Finally, the distance between source and receiver may be increased, or,

NOISE LEVEL INTERPOLATION AND MAPPING 339

if the receiver is an industrial worker, he may be fitted with hearing protection.

COMPLIANCE WITH FEDERAL AND MUNICIPAL CODES AND STANDARDS

Basically, the Occupational Safety and Health Administration (OSHA) permits 8 hours per day exposure to noise levels of 90 dBA. The exposure time limit is halved for each 5 dBA increase up to the maximum permitted continuous noise level of 115 dBA.[5] The allowable levels represent a compromise between actual existing noise levels in industry and safe levels in terms of hearing damage risk. Thus, OSHA levels may be subject to revision to reduce further the risk of hearing loss.

Community noise codes are generally based on annoyance and speech and sleep interference rather than hearing damage risk. Permitted levels for noise that intrudes on residential properties are much lower than permitted levels within factories. One city code,[6] for example, has different requirements for light and heavy manufacturing zones. The combined output of a corporation in a restricted manufacturing zone is not to exceed 55 dBA at the boundary of a residential zone. The limit is 61 dBA for general manufacturing zones; for general and heavy manufacturing zones, measurements may be made at the zone boundary or 125 feet from the nearest property line of a plant or operation, whichever distance is greater. One state code limits continuous noise from industrial operations to 65 dBA between 6 a.m. and 10 p.m. with a nighttime limit of 50 dBA.[7] For this code, measurements are made at a residential property line (even if the residence does not lie in a residential zone). The above noise limit values are given as illustrations only. The latest applicable codes and regulations must be obtained in every case.

Statistical (percent exceeded) noise levels are sometimes used to describe noise that varies with time. The most commonly used levels are L_{10}, L_{50} and L_{90} values expressed in dBA and based on one hour measurement periods. The L_{10} level is exceeded 10% of the time. For example, $L_{10} = 65$ dBA indicates that a noise level of 65 dBA was exceeded for 6 minutes during a given hour. Noise level L_{50} is the median and L_{90} is the level exceeded 90% of a given time period.

NOISE LEVEL MAPPING

The distribution of resources applied to noise control should be based on careful prediction, measurement and planning. The plotting of noise contours (points of constant noise level) can be an important step in noise control. Using a few predicted or measured noise level values, it is possible to estimate and plot noise levels at other locations. This permits application of engineering controls to reduce noise transmission paths. Noise plots may also aid in siting of noise-producing operations to reduce noise impact on employees and on the neighboring community. Other administrative controls including stationing of employees and the requirements for personal hearing protection may be indicated by noise level plots.

POINT SOURCES OF NOISE

A single piece of machinery, an exhaust stack, a fan or other relatively concentrated noise source may be considered a point source. At locations sufficiently distant from the source, noise intensity follows the inverse square law, providing reflections are not present.

THE INVERSE SQUARE LAW

In the absence of barriers, an ideal point source generates a spherical wave. Noise intensity (watts/m^2) decreases with distance according to the inverse square law:

$$\frac{I_2}{I_1} = \frac{r_1^2}{r_2^2} \qquad (18.3)$$

where intensity I_1 is measured at distance r_1 from the source and intensity I_2 at distance r_2.

Combining Equations 18.2 and 18.3, we obtain the effect of distance on sound level L(dBA) for an ideal point source:

$$L_2 = L_1 - 20 \log_{10}(r_2/r_1) \qquad (18.4)$$

where L_1 and L_2 refer to sound levels at distances r_1 and r_2, respectively. When dealing with airborne sound under ordinary conditions, there is no need to distinguish between sound *pressure* level and sound *intensity* level. It can be seen that sound level due to an ideal point source decreases by about 6 dBA per doubling of distance from the source. This relationship holds for a point noise source on a hard floor as well.

THE NEAR AND REVERBERANT FIELD

No actual machine can be exactly represented by an ideal point source. The region close to an actual noise source is called the near field. Noise levels fluctuate within the near field, and the 6 dBA sound decrease per doubling of distance law (inverse square law) does not hold. The extent of the near field will be of the same order of magnitude as a characteristic dimension of the noise source. The far field is the region in which sound level decreases by 6 dBA per doubling of distance. Smooth walls that reflect sound waves are called "acoustically hard." In a room, the sound level may fluctuate near walls. The region near a wall in which the 6 dBA per doubling of distance law does not hold is called the reverberant field. The near field, far field and reverberant field are illustrated in Figure 18.1.

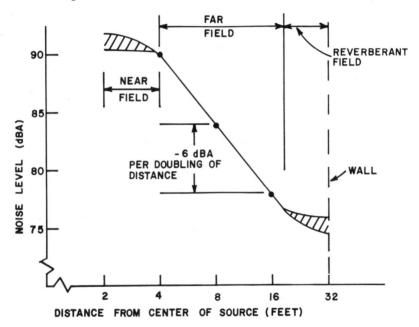

Figure 18.1 An illustration of the variation of noise level with distance from a point source (semilog coordinates).

Noise level is also influenced by barriers, by atmospheric attenuation and by background noise. Barriers that interrupt (or nearly interrupt) the line of sight between noise source and receiver cause a substantial reduction in noise level. Atmospheric attenuation causes additional noise reduction when sound is transmitted for hundreds of feet. Background noise (traffic noise and noise from sources other than the machine in question) becomes prominent as distance from the source increases.

PLOTTING POINT SOURCE NOISE LEVELS

Using Equation 18.4, the difference in noise level ΔL(dBA) measured at distances r_1 and r_2 from a point source is given by:

$$\Delta L = L_2 - L_1 = -20 \log_{10} (r_2/r_1) \tag{18.5}$$

Conversely, for a given difference in noise level ΔL, the ratio of distances to a point source is given by:

$$r_2/r_1 = 10^{-\Delta L/20} \tag{18.6}$$

An increase in noise level, of course, corresponds to a decrease in distance. Using Equation 18.6, the relationship between changes in noise level and the ratio of distances from a point source is given in Table 18.1.

Table 18.1 Change in Noise Level with Distance from a Point Source

Change in Noise Level ΔL (dBA)	Distance Ratio r_2/r_1
0	1.000
1	0.891
2	0.794
3	0.708
4	0.631
5	0.562
6	0.501
7	0.447
8	0.398
9	0.355
10	0.316
12	0.251
20	0.100

An interpolator for point source noise may be constructed as shown in Figure 18.2. A vertical line of any convenient length is drawn at the left of the figure. The top of the line is marked 0 dBA and the bottom of the line represents the noise source location. Measuring from the

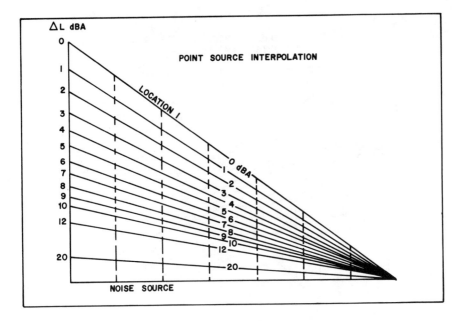

Figure 18.2 Point source noise interpolator.

bottom of the line, 0.891 times the length of the line corresponds to $\Delta L = 1$ dBA; 0.794 times the length of the line corresponds to $\Delta L = 2$ dBA, etc., as given in Table 18.1. Lines are drawn from each of these points to a point of convergence at some convenient location at the right of the figure. The interpolator of Figure 18.2 may be duplicated directly on a sheet of clear plastic, or a larger interpolator may be constructed from the data of Table 18.1 using transparency marking pens on clear plastic.

If the noise level is known at one point, the interpolator may be used to estimate noise level at another point *in line with the first point and the noise source.* Using a scale drawing of the area, the interpolator is oriented so that the "location 1" line lies over the point farther from the source and the "noise source" line lies over the source. The line lying over the nearer point indicates the difference in noise level between the two points.

Figure 18.3 illustrates the use of the interpolator when the noise level is unknown at a point between the source and a point of known noise level. In this example, the interpolator gives $\Delta L = 3$ dBA, which is added to the known value of 70 dBA for a noise level of 73 dBA at the point

344 INDUSTRIAL NOISE CONTROL HANDBOOK

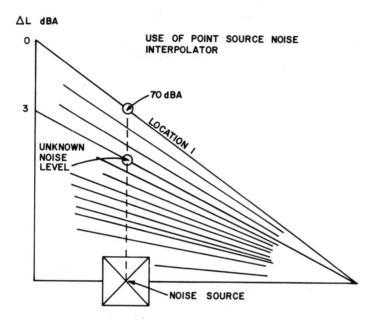

Figure 18.3 Use of point source noise interpolator.

nearer to the source. If the point of unknown noise level is farther from the source than the known point, ΔL is subtracted. If one of the points is very close to the noise source, it may lie in the near field, making the interpolation invalid. If one of the points is very far from the noise source, noise from sources other than the source in question may have a predominant effect.

LINE SOURCES OF NOISE

When a number of pieces of machinery or other noise sources of equal sound power lie in a line, the situation may be approximated by a line source. At points sufficiently distant from the line source, noise intensity follows the inverse distance law, providing reflections are not present.

THE INVERSE DISTANCE LAW

In the absence of barriers, an ideal line source generates a cylindrical wave. Noise intensity decreases with distance according to the inverse distance law:

$$\frac{I_2}{I_1} = \frac{r_1}{r_2} \tag{18.7}$$

In terms of sound levels L_1 and L_2 (dBA) at distances r_1 and r_2, respectively, this becomes:

$$L_2 = L_1 - 10 \log_{10}(r_2/r_1) \tag{18.8}$$

Thus, the sound level due to an ideal line source decreases by about 3 dBA per doubling of distance from the source.

PLOTTING LINE SOURCE NOISE LEVELS

Using Equation 18.8 for a given difference in noise levels, ΔL, the ratio of distances to a line source is given by:

$$r_2/r_1 = 10^{-\Delta L/10} \tag{18.9}$$

Using Equation 18.9, the relationship between changes in noise level and the ratio of distances from a line source is given in Table 18.2.

Table 18.2 Change in Noise Level with Distance from a Line Source

Change in Noise Level ΔL (dBA)	Distance Ratio r_2/r_1
0	1.000
1	0.794
2	0.631
3	0.501
4	0.398
5	0.316
6	0.251
7	0.200
8	0.158
9	0.126
10	0.100

An interpolator for line source noise may be constructed as shown in Figure 18.4, using the data from Table 18.2. If the noise level is known at one point, the interpolator may be used to estimate noise level at another point if both points lie on a perpendicular to the line source.

346 INDUSTRIAL NOISE CONTROL HANDBOOK

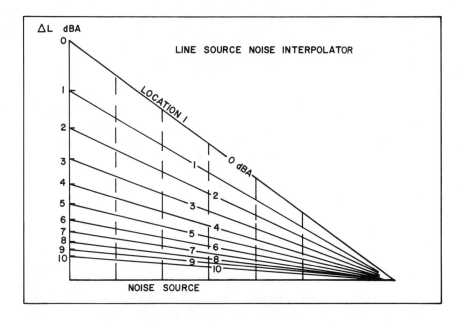

Figure 18.4   Line source noise interpolator.

The interpolator is oriented so that the "noise source" line lies over the source and the "location 1" line over the farther point. The line lying over the nearer point indicates the difference in noise level between the two points.

COMBINED NOISE SOURCES

When two or more independent noise sources are present, the effect at a given point may be obtained by adding the noise intensities (watts/m^2). Most measurements and predictions, however, are given as noise levels L (dBA). For separate noise level contributions L_1 and L_2, the total noise level L_T dBA at a point is:

$$L_T = 10 \; \log_{10}(10^{L_1/10} + 10^{L_2/10}) \qquad (18.10)$$

It can be seen that for $L_1 = L_2$, $L_T = L_1 + 3$ dBA (approx.). In general, if L_1 is larger than L_2 by N dBA, the combined effect of L_1 and L_2 is larger than L_1 by:

$$L_T - L_1 = 10 \; \log_{10} \; (1+10^{-N/10}) \qquad (18.11)$$

NOISE LEVEL INTERPOLATION AND MAPPING

The difference is given in Table 18.3 for convenience in combining noise levels. The results are given in chart form in Figure 18.5.

Table 18.3 Combined Noise Levels

Difference in Noise Levels to be combined (L_1-L_2 dBA)	Amount Added to Larger Level to Obtain Total Noise (L_T-L_1 dBA)
0	3.01
1	2.54
2	2.12
3	1.76
4	1.46
5	1.19
6	0.97
7	0.79
8	0.64
9	0.51
10	0.41
15	0.14
20	0.04

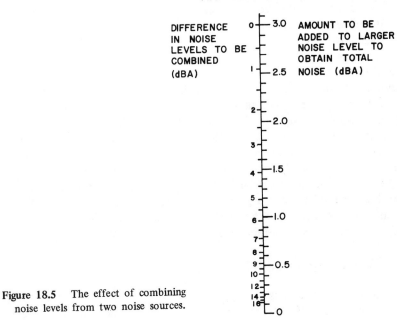

Figure 18.5 The effect of combining noise levels from two noise sources.

Using the table or chart, any number of noise contributions can be combined, two at a time. For example, let three machines produce levels of 70 dBA, 70 dBA and 67 dBA at a given point when operating singly. The effect of the three operating together is given by combining the first two levels to obtain 73 dBA, and then combining 73 dBA and 67 dBA to obtain a total noise level of 74 dBA (approximately). Due to the difficulty in making precise noise measurements and the tendency of noise levels to vary with time, final results are generally rounded off to whole decibels. It can be seen that if two noise contributions are to be combined and one is 15 or 20 dBA below the other, the lower noise level will not have a significant effect. For example, 80 dBA combined with 60 dBA produces a total noise level of approximately 80 dBA.

NOISE CONTOUR MAPPING FOR LINE AND POINT SOURCES

It is sometimes necessary to map noise contours to estimate the impact of industrial noise on workers or on the neighboring community. The noise interpolators may be used as an aid in plotting in a free field, sufficiently far from the noise source or barriers or reflecting surfaces. For an ideal line source, constant noise level contours are simply lines parallel to the source. For an ideal point source, constant noise level contours are concentric circles. For a source that shows some directionality, the contours will be distorted accordingly.

Figure 18.6 illustrates a plan for **installing** a row of machines that may be approximated by a line source. Based on measurements on similar machines, it is estimated that the row of machines will produce a noise level of 78 dBA at the points indicated. The line source interpolator was used to draw lines of constant noise level parallel to the source as shown in the figure. A single machine will also be installed, and it will be approximated by a point source. Measurements indicate that it will produce a noise level of 80 dBA at the points indicated. The point source noise interpolator was used to estimate the location of curves of constant noise level due to the single machine. No curves are drawn near the machines since the point source and line source idealizations are not valid in the near field. Barriers, reflecting walls and ceilings must also be considered. In this example, it is assumed that the area of interest is free of obstructions.

NOISE CONTOURS FOR COMBINED NOISE SOURCES

Figure 18.6 illustrates the approximate effect of two noise sources operating separately. If both sources operate together, the combined effect will be of interest. This may be obtained as in Figure 18.7, considering

NOISE LEVEL INTERPOLATION AND MAPPING   349

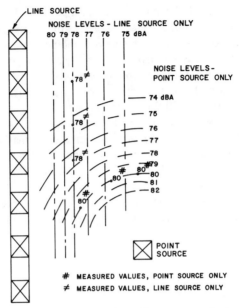

Figure 18.6   Noise contours for a line source and a point source.

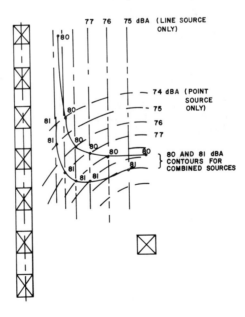

Figure 18.7   Noise contours for combined sources.

the intersection points of the noise contours and using Table 18.3 (or Figure 18.5) to sum the noise contributions. For example, the intersection of two 77 dBA contours represents a combined effect of 77 + 3 = 80 dBA, and that point is so marked. The intersection of a 74 dBA contour and an 80 dBA contour represents a combined effect of 80 + 1 = 81 dBA (approximately) and it is so marked. As indicated previously, precision of better than 1 dBA may not be warranted.

The combined noise contours are obtained by joining all of the intersections marked 80 dBA, then all of the intersections marked 81 dBA, and so on. In the illustration, only the 80 and 81 dBA combined noise contours have been plotted. If noise levels are of interest at greater distances from the noise sources shown, the plotted values may approach background noise levels. In that case, the background noise levels would be combined with the noise levels due to the point and line sources.

REFERENCES

1. Glorig, A. "Damage-Risk Criteria for Hearing," *Noise and Vibration Control*, L. Beranek, Ed. (New York: McGraw-Hill, 1971).
2. Kryter, K. "Impairment of Hearing from Exposure to Noise," *J. Acoustical Soc. Amer.* 53(5):1211-1234 (1973).
3. U.S. Dept. of Housing and Urban Development. *Noise Abatement and Control.* Circular 1390.2 (August 4, 1971).
4. Kavagodina, I. L. "Hygenic Importance of the Problem of Noise Abatement in the Cities," *Soviet Noise Research Literature.* U.S. EPA-55019-74-002 (April 1974).
5. Department of Labor. "Occupational Safety and Health Standards," *Federal Register* 37(202), Part II (October 18, 1972).
6. *Municipal Code of Chicago.* Chapter 17, 17-4.1 *et seq.,* effective July 1, 1971.
7. "Noise Control Regulations," *N. J. Administrative Code* 7:29-1.1 *et seq.* enacted January 23, 1974 (the nighttime limit became effective January 1, 1976), Chapter 29.

CHAPTER 19

GLOSSARY

Absorption—the ability of an acoustical material to absorb sound energy.

Acoustical material—a particular material that is engineered to absorb sound.

Acoustic output—the degree of a noise source in terms of sound power level or sound pressure level. The units are expressed in decibels.

Acoustics—the field of science that includes the study of sound generation and transmission.

Airborne sound—the sound that reaches the point of measurement traveling through the air.

Alias—Sampled data that is equally spaced.

Ambient noise—noise that is a mixture of different airborne sounds from many sources around the point of measurement.

Amplitude density distribution—the function expressing the fraction of time that pressure, voltage, or another variable dwells in a narrow range.

Analyzer—the combination of a filter system and a device used for evaluating the relative energy that passes through the filter system.

Anechoic—an environment that is very close to a free field.

Anechoic room—a room whose structure absorbs all incident sound. The purpose of such a room is to set up free field conditions.

Articulation index—an index aiding in properly classifying certain vowel and consonant sounds. It is labeled from 0 to 1 and is a fundamental method of identifying human speech sounds.

Audiogram—a plot of data that indicates the hearing threshold level (HTL) as a function of frequency.

Audiometer—a testing device used for measuring hearing ability and threshold levels.

Aural—that which has to do with the ear or to the sense of hearing.

Autospectrum—also known as the power spectrum. It is the spectrum whose coefficients of components are presented as the square of the magnitudes.

Background noise—noise that is picked up in testing that does not originate from the source of interest.

Baffles—components inserted in the flow path through rectangular silencers, usually to attenuate high-frequency noise.

Center bodies—elements inserted in the flow path through circular silencers, usually to attenuate high-frequency noise.

Coherence—a measure of a transfer function estimate. It has a value of one when the estimate has not been infiltrated by background noise.

Confidence limits—the limits (upper and lower) of a certain range over which a definite percent probability applies.

Critical speed—the speed of a rotating system that applies to a resonance frequency of the system.

Cross correlation—the measurement of similar characteristics of two functions whose displacement is an independent variable, usually time.

Cross spectrum—the measure inside the frequency domain of the similar characteristics of two functions.

Damp—to produce a loss of vibration or oscillatory energy in a mechanical or electrical system.

Damping—the loss of vibrational or oscillatory energy from a mechanical or electrical system.

Data window—the form of a weighing function that is considered as multiplying the data that enters a calculator.

Dead room—a room whose structural materials absorb most or all of any incident sound.

Decibel—a unit of sound pressure-squared level.

Degrees of freedom—a degree of stability related to the number of independent equal terms entering into a distribution.

Diffraction—any change in the direction of propogation of sound energy waves. Diffraction usually occurs at a boundary discontinuity of an absorptive material.

Diffuse sound field—an area in which the average rate of sound energy flow from a source is equal in all directions away from the source.

Directivity factor—the ratio of the square of the sound pressure to the mean square pressure. Used for sound emission in a transducer, this ratio is determined at a specified distance and direction from the source. For an accurate factor the distance from the source must be large so the sound will appear to diverge spherically.

Direct sound field—an area in which a majority of the sound comes directly from the source without any contributions due to reflection.

Displacement—a vector quantity indicating the change of position of a particle, measured from the mean position.

Dissipative silencers—an arbitrary designation for silencers that absorb rather than reflect noise, usually in a broad frequency range.

Dynamic insertion loss—insertion loss measured under flow conditions in silencers.

Earphone—an electroacoustic transducer that is tightly coupled to the ear.

Effective sound pressure—the root mean-square of the instantaneous sound pressures over a given length of time at the measurement point.

Face area—the cross-sectional area of the inlet portion of a silencer.

Face velocity—the velocity of a cross section of gas or liquid at the inlet face of a silencer.

Filter—a device separating the components of an incoming signal by its frequencies.

Flanking transmission—the transmission of sound from a source through a barrier that is not an acoustical material being directly tested. Such a barrier would be a partition.

Folding frequency—the inverse of twice the time interval between sampled values.

Frame—a group of values analyzed as a group.

Frame size—the number of sample points that are in the frame.

Free-sound field—an area in which the effects of the boundaries upon the sound field are irrelevant; a field that is free of boundaries.

Frequency—the rate of time for a full cycle to be completed in a periodic function. The units of frequency are Hertz (Hz).

Gaussian distribution—an amplitude distribution whose histogram is bell-shaped. Also known as normal distribution, it pertains in the sound field to stationary acoustic noise that is not periodic.

Hanning—a data window that is used in the form of a time domain of a raised cosine arch. At the beginning and end of the frame its weight is zero; in the middle, it weighs unity.

Hearing threshold level—the amount (in decibels) of the threshold of audibility of the ear exceeding the standard audiometric threshold.

Histogram—a plot of amplitude density distribution.

Impact—the collision of a moving mass into another mass that is at rest or also in motion.

Insertion loss—the difference in decibels between two sound levels that are measured at the same point in space before and after a muffler or a silencer is inserted between the measurement point and the source of the noise.

Instrument noise—the electrical sound that is generated in the measuring devices of airborne sound.

Isolation—the complement of transmissibility in a steady-state forced vibration.

Jerk—the third derivative of the displacement with respect to time. A vector giving the time rate of change of acceleration.

Just noticeable difference—the amount of difference required so as to be classified that there has been a change in any attribute sound.

Level—the logarithm of the ratio of a quantity to a reference quantity of the same type.

$$L = \log_r (q/q_0)$$

where: L = level of quantity (units measured in $\log_r r$)
r = reference ratio and the logarithm base
q = the quantity under consideration
q_0 = the reference quantity of the same field

Level distribution—an array of quantities that shows noise exposure by giving the duration of time that the sound-pressure level prevailed within each of a set of level intervals.

Line component—a simple tone. It may or may not be a portion of a complex signal.

Live room—a room whose sound absorptive qualities are very poor.

Loudness—a strong auditory sensation. Loudness is dependent upon sound pressure, frequency and wave form of the stimulating sound.

Loudness contour—a plot of data that shows related values of sound pressure levels and frequency that is needed to produce a specific loudness sensation.

Loudness level—the sound that is equivalent to the median sound pressure level relative to 0.0002 microbar of several trials evaluated by persons to be of equal loudness at 1000 Hz. Units are phons.

Loudspeaker—an electrostatic transducer that emanates acoustic power into the surrounding environment. The resultant acoustic waveform is equal to the electrical input.

Mach number—the ratio of the flow speed to the speed of sound in a silencer.

Masking—the magnitude that the threshold of audibility of a noise is increased by the presence of another masking sound within the environment. Units are expressed in decibels.

Mechanical impedance—the impedance obtained from the ratio of force to velocity during simple harmonic motion.

Mel—a unit of pitch. The pitch of a sound that is evaluated to be n times that of a one-mel tone is n mels.

Microbar—a unit of pressure used in acoustics. It is a dyne per square centimeter.

Microphone—an electrostatic transducer that is stimulated by sound waves, transmitting equivalent electric waves.

Muffler—a different name for a silencer.

Neper—a division on a logarithmic scale denoting the ratio of two similar quantities that are proportional to power or energy. This ratio is converted to nepers by multiplying the logarithm to the base e by ½.

Noise level—the actual quantity of noise that is measured in specific units such as decibels or volts.

Noise level for airborne sound—the sound level that is the weighted sound pressure level.

Noise reduction—a difference expressed in decibels between space-time average sound pressure levels that are produced in two rooms by any number of sound sources contained in them.

Noise reduction coefficient (NRC)—the average of sound absorption coefficients at frequencies of 250, 500, 1000 and 2000 Hertz.

Noys—a dimension used to evaluate the perceived noise level.

Octave—either the pitch interval between two tones or also the interval between two sounds with a frequency ratio of two.

Octave band—the frequency band whose upper band-edge frequency is double that of the lower band-edge frequency.

One-third octave band—the frequency band that has its upper band-edge frequency at $1/3$ the lower band-edge frequency.

Oscillation—the variation with time of the magnitude of a value with respect to a specific reference, when the magnitude is alternated greater and smaller than the reference.

Parallel silencer—a combination of silencers connected in parallel, such as putting one inside the other.

Partition—any structural component that separates space, such as wall, door, floor, or ceiling.

Peak-to-peak value—the algebraic difference between the two extremes of quantity of an oscillating parameter.

Perceived noise level—the level of noise that is obtained through a calculation process based on an approximation to subjective evaluations of noise magnitude (units are in decibels).

Periodic quantity—an oscillating quantity repeated over certain independent variable increments.

Phon—units used in denoting loudness levels.

Pink noise—a certain type of sound where the noise-power-per-unit-frequency is inversely related to the frequency over a specific range.

Pitch—the characteristic of an auditory sensation in which noises are ordered on a scale that extends from low to high.

Point source—a noise source that sets up a uniform sound field under free field conditions.

Power level—ten times the logarithm to the base 10 of the ratio of a certain power to a referenced power. Reference power is shown when the level is recorded (units are in decibels).

Presbycusis—loss of hearing due mainly to aging.

Pressure drop across a silencer—the difference between upstream pressure and downstream pressure at a certain flow rate through the silencer, normally measured in inches of water.

Pure tone—sound sensation that is characterized by singleness of pitch. Also referred to as a simple tone.

Random noise—an oscillation whose instantaneous magnitude cannot be specified for any point in time.

Rate of decay—the rate at which the sound pressure level decreases after a source has stopped giving off sound energy.

Reactive silencer—an arbitrary designation for silencers that are tuned to reflect, rather than absorb, noise in narrow frequency ranges.

Reflective environment—an environment in which large sound absorptive surfaces are present.

Resonance frequency—the frequency at which resonance takes place.

Reverberation—unyielding sound in an enclosed area as a result of repeated reflection after the source has stopped giving off sound energy.

Reverberation room—a room designed to approximate the reverberant sound field as a diffuse sound field.

Reverberant sound field—an enclosed area where most or all of the present sound is constantly reflected off its perimeters.

Salum—a unit of sound absorption measure.

Self-noise—the noise generated by air flowing through a silencer.

Series silencer—a combination of silencers connected in series, such as one located downstream from the other.

Side branches—sound-absorbing elements connected on the sides of silencers, normally for the purpose of attenuating low frequency noise.

Silencer—a device that allows a functional work flow while preventing the transmission of noise.

Sociocusis—a specific condition that increases the threshold hearing level due to exposure to noise that is not related to the social environment exclusive of occupational-noise exposure, physiologic changes over time, and otologic diseases.

Sone—the tiniest increment of a psychophysical scale indicating the intensive attribute of complex sounds. A unit of loudness.

Sonics—the science of sound in analysis and processing.

Sound—auditory sensation developed by the oscillations in pressure, stress, particle velocity and particle displacement in a medium containing internal forces, or the superposition of such propogated alterations.

Sound absorption—a process by which one eliminates sound energy.

Sound energy—energy that enters a medium where sound travels. It consists of potential energy in the form of deviations from static pressure and kinetic energy in the form of particle velocity.

Sound energy density—the amount of sound energy per unit of volume.

Sound intensity—the average rate of sound energy flow in a specific direction divided by the area, perpendicular to that specific direction, through or toward which it travels.

Sound power—symbolized by W and having the units of watts, it is the rate at which acoustic energy is radiated.

Sound pressure—a fluctuating pressure that is superimposed on the static atmospheric pressure. It is denoted by P and has the units of dyne/cm^2 or Newton/m^2.

Sound transmission class (STC)—a rating system using single figures that is engineered to give a preliminary approximation of the sound insulation properties of a partition or preliminary rank of a series of partitions.

Sound trap—another name for a silencer.

Tone—an auditory sensation having pitch.

Transducer—a device that is actuated by waves from transmission systems, and in turn transfers waves to other transmission systems.

Vibration—oscillation where the quantity is a parameter that defines the motion of a mechanical system.

Vibration isolator—a support that isolates a system from any steady-state excitation.

Vibration meter—an instrument that can measure the displacement, velocity, or the acceleration of a vibrating body.

Volume flow velocity—the rate at which a specific volume flows through a silencer, usually cubic feet per minute.

Weighting—a specific frequency response given in a sound level meter.

INDEX

acoustical experimentation 153-154,161
acoustical foam 175-178, 180 *ff.*
  for aircraft 188-189
acoustical measurement 316-317
acoustical panels 178-179
acoustical windows 172-174
administrative control 2,27
air-conditioning system noise control 260,294-295
aircraft noise suppressor systems 188,211-215
amplitude distribution analyzers 322,323
ANSI 305-325
atmospheric attenuation 341
audiograms 13-14
audiometers 325-329,338
audiometric testing 325-331
A-weighted sound level (L_A) 298

ballistocardiogram 12
barriers
  See noise

caulking 67,89
cavitation, controlling 284-287
coincidence theory 145-147, 159-160
combined noise 346
communication problems 11,79-80
community action 338
complaints 31
complex noise 52-53
compressors 127,295

construction site noise 128-129
contesting OSHA citations 32
cork 62,241,246
critical frequency 148-149

dampers, adjustable 244-245
damping compound 198-201, 260-262
decibel 47-48, 228-230,338
  defintion of 47-48
Department of Labor Occupational Noise Standards 1-2
dosimeters 317-321,331-334
  portable 331-334
DYAD
  See damping compound
dynamic transmission loss 298-299

ear
  examination 81
  parts of 6-7
ear protectors 69 *ff.*
  insert-type 72,75-79
  muff-type 75-79
  requirements 71,80-82
employee responsibilities under OSHA 19,30
employer responsibilities under OSHA 18,29,34-35
enclosures for noise control 128-132
engineering controls 26,55 *ff.*

foam products 175,176,183

360 INDUSTRIAL NOISE CONTROL HANDBOOK

frequency analyzers   318-321
frequency weighting   51,337

glass
   double   156-163
   for noise reduction   143 *ff.*
   laminated   165-171
glazing   155,162-164
   multiple   163-164

harmonic motion   41,43,220,226
hazard abatement   33
hearing loss   2,338

impact (impulse) noise   25
insertion loss   298
inspections   31-32
interpolation   337,342-348
inverse square law   340
isolation
   of machine operators   65-66
   of noise sources   64-65

jet engine testing   204

laminate, natural frequency of   169-170
lead, as a noise barrier   83-85, 134-138
Long Island Lighting Company   113-117
loudness   50-51
   contour   50

machine guards   85
machine maintenance   55-57
machine substitution   26,57-58,158
mapping   309,337,339-340,348
microphones   47,305-311
Model Community Noise Ordinance   39
mounts for equipment   241-242

National Institute of Occupational Safety and Health   17,20-21
neoprene   246
noise
   effects on body   12-13
   effects on ear
     See ear
   legislation   17-39
   psychological effects of   10-12
   tolerance   7-8
   Also see complex noise, impact noise, variable noise
noise barriers   83,94-111
noise criteria (NC)   295-297
   preferred (PNC)   296-297
noise levels
   combined   346-350
   interpolation   337,342-348
   mapping   339,348
noise pollution level (NPL)   297-298
noise source isolation
   See isolation
nylon gears   189-194

Occupational Safety and Health Act (OSHA)   1,17-19,22*ff*, 315,317-320,324,325,339
octave filter   315,320

pads for equipment   241-242
path treatment   276-278
penalties for violation of OSHA   32,33
periodic functions   219-222
phon   50-51
piezoelectric accelerometer   234-235,237
piping systems   278-281
plenum barriers   60,85-93,302
   installation of   87-93
polyvinylchloride (PVC)   194-195

recordkeeping requirements   33, 34,36-37,328

INDEX 361

refinery noise reduction 118-119
resonance 149-151,159-160
root mean square values 47, 230-231
rubber isolators 242-243

shaft couplings 256
silencers 60,203 ff.
sound
  nature of 41 ff.
  propagation of 41 ff.
sound barriers 94-111
sound field 44-47
sound level meter 299-300,312-316,320
sound transmission class (STC) 94-112,152-153
sound transmission loss (STL) 151-152,155
spinning mill noise reduction 120-128
spring systems 243-244,263-270
strain gauge 232
suppressor systems 203 ff.
systems design 55,66-68

temperature effects on transmission loss 170-171
transducers 231-235
transmission loss (L_{TL}) 145,149 298,302,338

transmission loss curves 94-111

valves 273-275
valve noise, hydrodynamic control of 283-292
variable noise 25
variance from OSHA standards 33
ventilating system noise control 293-303
vibration
  causes of 60-61
  effects of 223-224
  effects on man 14-15
  measurement of 224-226
  nature of 219-223
  random 222
  sources of 223
vibration damping 59-62
vibration meter 235-240
vibration plates 257
vibrograph 231-232
vinyl 186
  coated fabric 68
  fiber-loaded 196
  lead-loaded 197-198
viscoelastic adhesive 84

wall treatment 187
Walsh-Healey Public Contracts Act 27-28
washer-dryer system of noise control 133-134
Workmen's Compensation 36